Titles in this series:

Geothermal energy

Review of research and development

Edited by H. Christopher H. Armstead

Second impression 1974

The Unesco Press
7 Place de Fontenoy, 75700 Paris

Composed by Imprimeries Réunies de Chambéry
Printed by Offset-Aubin, Poitiers

LC No 72-97138

The Unesco Press

First published 1973
Second impression 1974

The Unesco Press
7 Place de Fontenoy, 75700 Paris

Composed by Imprimeries Réunies de Chambéry
Printed by Offset-Aubin, Poitiers

ISBN 92–101063–8
LC No 72–97138

Preface

Although sources of geothermal energy have been known and exploited for many years, notably in Iceland, Italy and New Zealand, interest in the possibility of finding and developing new sources in other parts of the world has only recently become general. On two occasions the United Nations acted to coordinate the vast amount of geothermal experience, research, theorising and analysis that has been undertaken by scores of workers all over the world. The first occasion was the convening of the United Nations Conference on New Sources of Energy, held in Rome in 1961. A large part of the work of that conference, including the submission of eighty papers, was devoted to geothermal energy. The second occasion was nine years later, when a United Nations Symposium on the Development and Utilisation of Geothermal Resources took place at Pisa, Italy, in 1970. At this symposium, attended by nearly three hundred participants, 182 papers were submitted. Both the Rome and Pisa meetings were truly international in character: participants and authors were drawn from a wide range of countries. Much time was devoted on both occasions to discussions and the interchange of ideas.

Meanwhile, at its fourteenth session in 1966, the General Conference of Unesco adopted a resolution authorizing the Director-General to promote and facilitate international co-operation in the scientific study of the earth, including the creation and reinforcement of centres for research and training in techniques for the exploration of sources of geothermal energy.

In 1968, Unesco convened an Ad Hoc Working Group of experts selected from countries which had already acquired experience in this field, to advise on what steps might be taken to meet the ever-growing demand for trained specialists in the exploration of sources of geothermal energy—mainly geologists, geophysicists or geochemists and drilling or production engineers. With a view to stimulating the training in these fields, Unesco supported the organization of courses, notably at the University of Kyushu (Japan) and at the Centre for Geothermal Research in Pisa (Italy).

The present Review offers a general introduction to the subject of geothermal energy. The Review makes no claim to being an encyclopaedia of the geothermal arts: it is intended only to be an 'A-B-C' to enable the reader to gain an elementary insight into the various phases of geothermal work, from exploration to utilisation, including such related topics as earth structure, geothermal economics and field management. The authors are responsible for the opinions they have expressed in this book.

Unesco wishes to express its gratitude to Mr. H. Christopher H. Armstead who kindly accepted to act as the technical editor for this publication and to the authors for their valuable contributions towards the preparation of this Review.

Contents

Foreword

Exploration for geothermal energy and its subsequent exploitation, calls for the coordinated efforts of a team of specialists versed in different disciplines. Each specialist can gain a better perspective of his function in this teamwork if he has at least some understanding of the problems and techniques of his collaborators. This Review should enable him to do this: it should also enable the non-specialised reader to obtain a broad idea of geothermal energy in general and of its associated specialisations in particular.

The Review contains articles written by specialists. Some of the work of these specialists is so interwoven with that of their colleagues that it is impossible to confine each discipline within a 'water-tight compartment': a degree of overlap is unavoidable. For example, the geologist and the geophysicist must both rely on the work of the geochemist to such an extent that neither can describe his work adequately without some mention of geochemistry. Repetition,

however, has been here limited to ensuring that each article can stand by itself as a self-sufficient exposition of its subject.

Those readers who become sufficiently interested in the subject of an article to wish to pursue their quest for knowledge further are referred to the extensive bibliographies which appear at the end of nearly all the articles. Many of the references in turn quote further bibliographies, so that the reader's studies may be guided and extended as far as he wishes.

Even today, there are people who are under the impression that there is something freakish and impracticable about geothermal energy. This Review should correct that impression and persuade the reader that the economic exploitation of geothermal energy is no longer a dream, but one of the established facts of life.

UNIT CONVERSION TABLE

As the various authors of the articles in this Review have not all adopted the same system of units, the following conversions may be helpful:

Class	To convert...	into...	multiply by...
LENGTH	centimetres (cm)	inches (in)	0.3937
	metres (m)	feet (ft)	3.281
	metres (m)	yards (yd)	1.094
	kilometres (km)	miles	0.6214
AREA	sq. centimetres	sq. inches	0.155
	sq. centimetres	sq. feet	0.001076
	sq. kilometres	sq. miles	0.3861

Class	To convert...	into...	multiply by...
VOLUME	cu. centimetres	cu. inches	0.06102
	cu. metres	cu. feet	35.31
	cu. metres	litres	1,000
	cu. metres	U.S. gallons	264.2
	cu. metres	Imp. gallons	220.5
	U.S. gallons	Imp. gallons	0.833[1]
	U.S. gallons	cu. ft	0.1337
MASS	kilograms (kg)	pounds (lb)	2.205
	kilograms	long tons	0.0009842[2]
	kilograms	metric tons (tonnes)	0.001
	kilograms	short tons	0.001102[3]
DENSITY	grams/cm^3	lb/in^3	62.43
	kilograms/m^3	lb/ft^3	0.06243
PRESSURE	atmospheres	cm of mercury	76
	atmospheres	kg/cm^2	1.033
	atmospheres	lb/sq. in (psi or lbf/in^2)	14.70
	bars	atmospheres	0.987
	bars	newtons/m^2	100,000
	bars	psi (lbf/in^2)	14.50
	bars	kg/cm^2	1.020
ENERGY	British thermal units (BTU)	ergs	1.055×10^{10}
	BTU	foot-pounds	778.3
	BTU	joules	1,054.8
	BTU	kg-calories	0.252
	BTU	kilowatt-hours (kWh)	0.0002928
CONCENTRATION	parts per million (ppm)	grains/U.S. gal	0.0584
	ppm	grains/Imp. gal	0.07016
CALORIFIC VALUE	k cal/kg	BTU/lb	1.8
ENTHALPY	joules/gram (J/g)	BTU/lb	0.430
HEAT FLOW	megawatts (MW)	BTU/hour	3,410,000

TEMPERATURE CONVERSION

Degrees Fahrenheit (°F) = 1.8 × degrees centigrade (°C) + 32.
Degrees Centigrade (°C) = (degrees Fahrenheit (°F) — 32) × 5/9.
Absolute Centigrade temperature = °C + 273.
Absolute Fahrenheit temperature = °F + 460.

PREFIXES

micro (μ) = one millionth
milli (m) = one thousandth
centi (c) = one hundredth
deci (d) = one tenth
deka (D) = ten times

hecta (h) = one hundred times
kilo (k) = one thousand times
mega (M) = one million times
giga (G) = 10^9 times
tera (T) = 10^{12} times

1. 1 Imperial gallon of water at 4 °C weighs 10 lb.
   1 U.S. gallon of water at 4 °C weighs 8.33 lb.
2. 1 long ton = 2,240 lb.
3. 1 short ton = 2,000 lb.

I General

What is geothermal energy?

H. Christopher H. Armstead

Consulting Engineer
Rock House, Ridge Hill,
Dartmouth, South Devon
(United Kingdom)

The nature and occurrence of geothermal energy

There was a time, not so long ago, when geothermal energy was regarded merely as an interesting freak of nature, a tourist attraction in the form of geysers, fumaroles and pools of boiling mud. Its practical side was then more or less confined to its alleged curing properties for various human ailments and to the availability in somewhat remote places of 'natural' hot baths. Those days have passed, and underground heat has become an undisputed commercial competitor with other forms of energy for many applications.

Although we are rapidly becoming familiar with the conditions of outer space, we have, until recently, been surprisingly ignorant of what is going on only a mile or two beneath our feet. As long ago as 1904 the British engineer, Sir Charles Parsons, sought to remedy some of this ignorance by advocating the sinking of a shaft 12 miles deep into the earth. He estimated that this would then have cost £5 millions—a very considerable sum in those days— and would have taken 85 years to construct. This proposal, light heartedly referred to by its author as 'the Hellfire Exploration Project', was never implemented.

As everyone knows, the nether regions of the earth are very hot, but in most parts of the world the observed temperature gradient in the outer crust averages only about one degree centigrade for every hundred feet of depth. On this basis Sir Charles Parsons expected to encounter temperatures of the order of 600 °C, or 1,100 °F, at the base of his shaft.

In certain regions of the earth much steeper temperature gradients occur—sometimes as much as a hundred times the normal: it is the heat in these regions that is termed 'geothermal energy'. Such 'thermal' regions are usually, but not always, closely associated with volcanic activity and earthquakes. A glance at Figure 9 of the article by E. Bullard will show the distribution of the world's principal earthquake zones. The dots and crosses in that map represent actual recorded earthquakes (shallow and deep, respectively) during the years 1960-67. The vast majority of these earthquakes occurred in clearly defined belts, or zones, mostly of comparatively narrow width. The most important of these earthquakes zones, which also contains a great number of active and extinct volcanoes, more or less follows the periphery of the Pacific Ocean: this zone is sometimes known as 'the belt of fire'. Another important zone runs along the middle of the Atlantic Ocean with an easterly branch passing through the Mediterranean and the Middle East into Tibet. Yet another important zone more or less follows the direction of the Great African Rift and the Red Sea. All these zones are interconnected with one another, except for a small isolated zone centred on the Hawaian Islands.

Geothermal areas generally tend to lie within these earthquake belts, though not necessarily close to volcanoes. For instance, Larderello in Italy and The Geysers in California, two of the most famous geothermal fields, lie at considerable distances from the nearest volcanoes. Thermal areas are also known which occur outside the earthquake zones—e.g. Kenya, Hungary and various parts of the Soviet Union. Nevertheless, all the more 'spectacular' thermal areas are found within the earthquake zones, which can accordingly be regarded as containing the most promising areas for geothermal exploration, though they hold no 'monopoly' of such areas. It is unfortunate that such a large proportion of the earthquake belts lies beneath the seas.

Various theories and 'models' have been advanced to account for the origin of geothermal heat and the mechanism of its upward transfer towards the surface of the earth. There is still much speculation about what is actually happening below ground, but gradually the fog of uncertainty is lifting. The actual source of the heat is now almost universally believed to be radioactivity, mostly within the crustal rocks, while the theory of 'continental drift' offers an interesting and plausible explanation of the broad confinement of earthquakes to certain clearly observable zones. It is in these zones that crustal weaknesses enable deep seated heat to rise nearer to the surface of the earth. There has for some time been a proposal, known as the 'Mohole Project', to drill about 7 miles through the crust of the earth where it is thinnest, namely in the bed of the deepest part of the ocean, in order to penetrate the Mohorovičić boundary between the earth's outer crust and the hot 'mantle' beneath. This operation, which has unfortunately been repeatedly postponed, could be regarded as a new and restricted 'Hellfire Project' which will probably cost 7 or 8 times as much as the original proposal of Sir Charles Parsons. It is hoped that the eventual execution of the Mohole Project may do much to reveal the true nature of the mantle and will greatly advance our knowledge of the 'working' of our planet.

For the present, the simple fact may be accepted that in certain parts of the world Nature has provided us with vast quantities of accessible geothermal heat.

History of geothermal exploitation

It is not surprising that Man has sought to make use of this heat. In ancient times the Romans, and in modern times the Icelanders, Japanese, Turks and others have used it for baths and for space heating. The Maoris in New Zealand too have exploited natural heat for their domestic needs. One of the more interesting sights in their country is a Maori village near Rotorua, in North Island, where one may see a fisherman catch his trout and drop it into a nearby pool of boiling water to cook it. A few yards away may be seen his wife administering a geothermal bath to the baby, while his daughter is doing the household laundry and the potatoes are cooking over a fumarole.

At Larderello in Tuscany the Italians have been extracting boric acid from steam jets ever since the 18th century and it was here that *power* was first successfully generated from geothermal heat. Even before the turn of the present century attempts were being made to use reciprocating engines supplied with raw natural steam. Now, a group of power stations in the Larderello district is collectively generating 390 MW. Great credit is due to the Italians for pioneering the development of geothermal power: their enterprise has been a source of encouragement to others.

The amount of power now being generated for public electricity supply purposes in various parts of the world from geothermal steam is over 700 MW (see Table). With plants now under construction this figure will rise to nearly 900 MW very shortly.

The New Zealanders are tentatively planning to develop a new field at Broadlands, while further substantial extensions are under consideration for California. The world's installed geothermal power capacity should very soon top the 1,000 MW figure.

Parallel with these power developments, geothermal energy has been extensively used for space heating and for industry.

Table of geothermal power generation

Installation	Installed in 1971	Under construction
Italy. Larderello group	390 MW	—
New Zealand. Wairakei	192 MW	—
California. The Geysers	82 MW	110 MW
Japan { Otake	13 MW	
Japan { Matsukawa	20 MW	
Iceland. Námafjall	3 MW	
U.S.S.R. Pauzhetka	5 MW	
Mexico. Mexicali	—	75 MW
	705 MW	185 MW
Total	890 MW	

Meanwhile, several other countries are taking an active interest in geothermal development. The Resources and Transport Division of the United Nations has given technical advice on geothermal development to no less than nineteen countries, while the United Nations Development Programme has in hand through its Special Fund, five geothermal exploration projects in Turkey, El Salvador, Chile, Kenya and Ethiopia. A request for a similar project has been submitted by the Government of the Philippines, and several other countries are becoming very much alive to the possibilities of geothermal development.

Applications of geothermal energy

From what has been said above it will be seen that the generation of power has hitherto been the most important application of geothermal energy. This is understandable when it is realised that geothermal fields are liable to occur at remote locations. The ability to transmit electric power over long distances to some extent immunises geothermal energy against these disadvantages of location, as power may be generated at the field and transmitted to far-off centres of population. Moreover, electrical energy is a readily marketable commodity, and a geothermal power plant can provide an extremely cheap and reliable supply of base load at very high load factor. Its availability, day and night, independently of rainfall and season, renders it in some ways more reliable than its principal rival—hydro power.

It is probable that power generation will remain the most important application for geothermal energy in future. Nevertheless this form of energy is far too versatile to be restricted in use to a single application. Already it is being extensively used for district heating, particularly in Iceland, and for raising 'forced' vegetables and flowers in heated glass greenhouses in Iceland and in parts of the U.S.S.R., where the climate is too harsh to support normal husbandry. It is also being used for heat-intensive industries—for example, for paper making in New Zealand and for the recovery and processing of diatomite in Iceland. Air conditioning is being successfully performed in New Zealand by geothermal means and borax is still being produced geothermally at Larderello in Italy, in continuation of an industry established in the 18th century.

However, the relatively few applications that have actually been established do not reflect the immense potential of geothermal energy for industrial and other purposes. Some possible developments in this direction are mentioned in articles by S. Einarsson and H. C. H. Armstead of this Review.

Perhaps the most pressing human need in the coming years will be fresh water. Geothermal desalination holds out great promise of cheap and abundant supplies of fresh water at certain places, though the choice of such places is admittedly limited by restrictions of field location.

Some geothermal waters are known to contain small proportions of highly valuable mineral ingredients, and their extraction holds out considerable promise. The low concentrations of these ingredients can be offset by the high yields of the containing waters. Great interest is now being shown in the Salton Sea area in California, where hot geothermal brines are known to contain precious metals.

The industrial and other potentialities of geothermal energy suggest that great economic advantages could be gained from dual or multi-purpose plants combining power production with one or more other application. Such plants would enable the costs of exploration drilling and certain other items to be shared between two or more end products.

While power may remain the dominant reason for exploiting geothermal fields, it is important that planners should not be blind to these other wide potentialities.

Obstacles to geothermal development

In view of the cheapness of geothermal energy wherever it has been exploited, it may appear surprising that development has been slow and that so many geothermal fields are still neglected. The reason is not far to seek. Not all geothermal fields are necessarily amenable to economic exploitation, and in order to determine whether a field can be profitably put to use it is necessary to expend fairly formidable sums in carrying out exploration. If the results are positive, these exploration costs will have been well justified; but if negative, they will have been largely wasted except in the interests of pure science.

Thus the costs of geothermal exploration may be regarded as 'risk capital', such as is associated with petroleum exploration. But with petroleum exploration, if successful, the product can be shipped and sold all over the world regardless of the location of the oil field. In the case of a successful geothermal field, however, the energy must be used either locally or within a limited radius. The most probable market is electricity supply. But electricity supply can develop by conventional means without the need for 'risk capital'. As electricity is normally generated at no profit in the case of publicly owned enterprises, or at modest profit in the case of private enterprises, there is little incentive to expend 'risk capital' when development can take place along alternative, financially safer, lines.

The reluctance to embark upon fairly costly geothermal exploration projects is thus attribuable to this irreconcilability between the philosophies of 'risk capital' on the one hand and of 'safe development' on the other hand. Nevertheless, the economic advantages of exploiting a successful geothermal field can be so great that it is highly desirable that some means be found of bridging this philosophical gap.

It is in this respect that the United Nations Development Programme has rendered great service. The Special Fund activities of this programme are specially conceived to undertake pre-investment work and to relieve client governments of some of the risks which they feel they cannot afford. The five geothermal exploration projects referred to earlier in this article are examples of such work undertaken by the United Nations Development Programme. The results achieved in the more advanced projects are distinctly promising, and this fact gives encouragement to further exploration work. It is too early to judge the prospects of those exploration projects that have only recently been started.

Some of the more developed countries have felt confident to undertake their own geothermal exploration without the help of any international organisation. Thus Italy, New Zealand, California and Mexico have done their own work. The incentive for such action lies in the fact that many countries, having 'skimmed the cream' off their more attractive and easily exploitable energy resources, are becoming compelled to give greater consideration to less conventional sources of energy.

Geothermal prospects

The outputs of some hundreds of megawatts of geothermal power, referred to earlier in this article, amount of course only to a very small fraction of the world's total power production. Whilst the number of geothermal power plants is expected to rise fairly rapidly during the next decade or so, it must be admitted that in the foreseeable future the total power contribution from geothermal energy will remain a very small fraction of the total. Nevertheless, for certain developing countries the relative importance of geothermal power may be very impressive. Similar considerations apply to the use of geothermal energy for industrial and other purposes. Thus the short and medium term future for this form of energy, though important, should not be exaggerated.

As for the long term prospects of geothermal energy, these could be prodigious, for we have in the interior regions of the earth a store of energy so vast that others, which may conceivably lie within our grasp, pale into insignificance. The problem, of course, is to tap it, since Nature allows only such a tiny fraction of it to leak through to near the surface. At present, geothermal exploitation is confined to 'thermal' areas where the downward temperature gradient is exceptionally high by comparison with the average of about one degree centigrade per hundred feet of depth. In non-thermal areas there are still immense reserves of heat below ground, but at such depths that they cannot be tapped economically by existing means. However, this will not necessarily always be so. It can be shown that if our planet were to be cooled through one degree centigrade, enough heat would be released to keep the whole world supplied with all its electrical power needs at their present level for some forty million years. It is not of course suggested that we should seriously set about cooling the earth by one degree, or even by one-hundredth of a degree: the

seismic and climatic consequences alone could be disastrous. But it is not inconceivable that in the course of time ways may be found of penetrating some distance into the magma, possibly with the help of underground nuclear explosions, at a commercially acceptable cost in terms of the prizes to be won. In this way we shall perhaps be able to tap at least a minute fraction of this vast store of energy in large local concentrations which, in combination with the waters of the oceans, could solve both the energy and fresh water problems of Mankind virtually for all time—furthermore with a minimum of associated pollution.

All this may be regarded as fantasy, but only a score of years ago space travel too was regarded as a fantasy. It is interesting to speculate whether Humanity might not have reaped a far greater reward had the vast expenditure hitherto lavished upon space exploration, been devoted instead to activities in a downward direction.

Basic theories

Sir Edward Bullard

Professor of Geophysics,
University of Cambridge (United Kingdom)

Introduction

In this chapter a brief account is given of the structure of
the earth and of the processes going on within, particularly
of the generation and transport of heat. In the last few
years there has been a major revolution in geology in which
new ideas about the history of the earth and the nature of
the motions in it have been developed and widely accepted.
Here, the views of the successful revolutionaries are adopted: those of the remnant defending the traditional position
cannot be explained at every point. Briefly the traditional
position is that there have been no major horizontal motions
of large blocks of the earth's crust and that the whole outer
part of the earth to a depth of 3,000 km is essentially a
solid undergoing only vertical motion; these views have
been persuasively and ably expounded by Belousov (1970)
and Jeffreys (1970). Here a different view is taken.

Crust, mantle and core

At the surface of the earth a great variety of rock is found;
sometimes sediments, such as clays, sandstones or limestones; sometimes ancient 'shields' composed largely of
granite; and sometimes lavas which have poured out from
volcanoes. The sediments are of great importance for
geology—they contain all the oil and a major part of the
mineral resources of the earth; they also contain the fossils
on the study of which our knowledge of the history of the
earth so greatly depends; for the purpose of this article,
however, they are not of great significance.

Beneath the sediments of the continents there lies a
'basement' composed predominantly of granite. Some granites have been formed by the freezing of molten material
and some by the metamorphism of sediments by heat and
by liquids and gases rising from below. Granites are the
result of complex thermal, mechanical and chemical processes and have a varied chemical and mineralogical composition. In general it may be said that they represent

material with a low melting point and a considerable proportion of free silica in the form of crystalline quartz. In
one way or another this material has been separated from
more basic rocks (i.e. material containing less quartz)
deeper in the earth. The more basic material often appears
near the surface as basalt dykes cutting the granites; (a
dyke is a vertical sheet of igneous rock that has forced its
way upward and solidified).

It is likely that the proportion of basalt increases with
depth and in some places there may be a layer of almost
pure basalt underlying the granite. The study of the elastic
waves from earthquakes shows that at a depth of about
35 km beneath the continents there is a boundary known
as the 'Mohorovičić discontinuity' or 'Moho', after its
discoverer Andrija Mohorovičić. At this discontinuity the
velocity of seismic waves suddenly increases. For compres-
sional waves the increase is from about 7 to 8.1 km/s. The
discontinuity represents the lower boundary of the granites
and basalts which constitutes the 'crust' of the earth. On
the continents the mountain ranges are composed of masses
of sediments which have been squeezed, heated and distorted and have been intruded by molten rocks from below.
The forms of the mountains as we see them are the result
of the carving of such masses by flowing water and moving
ice.

In the oceans everything is different. The ocean floor
is a new geological world. All the rocks are basalts; there
are no granites either beneath the sea floor or on islands
(except for a few 'continental fragments', such as the Sey-
chelles, and a few patches of granite formed by the separation of minerals from large masses of molten basalt, as
on Ascension Island). Beneath the basalts there is a Mohorovičić discontinuity just as there is under the continents,
but it lies only about 5 km beneath the ocean floor, that
is 10 km below the sea surface. The structure of the continental and oceanic crust is illustrated in Figure 1. The
mountains of the oceans are not carved from folded rocks
but are volcanoes; when they emerge above the sea surface
they are often capped by coral, but beneath the coral there
is always a volcanic core of basalt.

Beneath the crust lies the 'mantle'. For forces lasting for a few seconds, such as those concerned in the propagation of the waves from earthquakes, its material acts as a solid. It also behaves as a solid in the natural vibrations of the earth with periods of up to an hour, for tidal periods of 12 hours and for the periods of around a year involved in wobbles of the earth about its axis of rotation. For the much longer periods involved in geological processes it may be expected to 'creep' as do all solids at temperatures in excess of a few hundred degrees. The most direct evidence for its behaviour for long continued forces is the rise of previously glaciated areas, such as Scandinavia, after the removal of the ice. This rise takes a time of the order of 10,000 years and is compatible with the material of the upper mantle behaving like a fluid with a viscosity of about 10^{21} g/cm.s. There has been much discussion of the probable rheology of the upper mantle and particularly as to whether it possesses a finite 'strength' below which it does not flow. The view taken here is that for periods of up to a day it is a very 'good' solid which literally 'rings like a bell', and that for periods greater than 10,000 years it is a 'Newtonian fluid' which will move under indefinitely small stress differences; (for a discussion of the physics, see McKenzie, 1968 and for a contrary view see Jeffreys, 1970). Its behaviour for periods between a day and 10,000 years is doubtless complicated: at the lower end it is presumably a solid with some type of imperfect elasticity, and at the higher end a liquid whose effective viscosity varies with the stress.

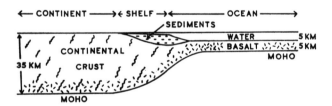

FIG. 1. Schematic section through earth's crust at an inactive coastline, such as that of eastern North America. Under the continent the proportion of basalt probably increases with depth, but it does not necessarily form a separate layer. A great thickness of sediments usually underlies the continental shelf.

The material of the mantle is not directly accessible to observation and our knowledge of it depends partly on deductions from the velocities of seismic waves and partly on the study of rocks at the surface that may be supposed to be derived more or less directly from it. Owing to the shallowness of the mantle in the oceans the study of rocks from the sea floor is of particular importance. The consensus of opinion is that at the Moho there is a marked change in composition and that the material below it is of ultrabasic composition, a peridotite consisting predominantly of the mineral olivine—iron magnesium orthosilicate, $(Mg, Fe)_2SiO_4$. The composition is probably quite similar to that of stoney meteorites.

With increasing depth in the mantle the pressure and temperature both increase and at some point changes in crystal structure may be expected. Seismology suggests a gradual change at depths in the neighbourhood of 400 km which probably corresponds to the changes towards more compact crystal structures that are observed at high pressures in the laboratory. It is possible that the viscosity increases greatly in the transition, but there is little direct evidence as to the amount of the change.

The mantle extends to a depth of about 2,900 km where there is a change to the much denser and liquid core. This is probably composed largely of molten iron. Within this liquid core is an 'inner core' of radius about 1,350 km which may be composed of iron that has been solidified by pressure.

The main facts of the large scale structure of the earth are summarised in Figure 2. For the purposes of this article only the crust and the upper part of the mantle are of importance. What lies below is too remote to affect the surface except on a very long time scale.

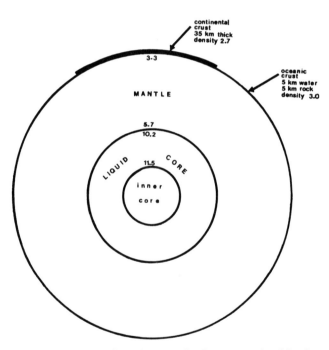

FIG. 2. Crust, mantle and core. The figures are densities in g/cm³.

Underground temperatures and heat flow

It has been known since the 17th century that the temperature in deep mines exceeds that at the surface of the earth. Such a temperature gradient implies a flow of heat outwards from the interior to the surface and raises many problems concerning the amount of heat involved, its variation from place to place and its origin. If the heat is

TABLE 1. Some typical measurements of heat flow

Place	Depth m	Greatest temperature °C	Mean gradient °C/km	Mean conductivity 10^{-3} cal/s cm °C	Heat flow 10^{-6} cal/cm²s
Gerhardminnebron (South Africa)	3,022	46.4	9.5	13.5	1.28
Tocketts (Yorkshire, England)	906	35.7	27.7	4.26	1.18
Adams Tunnel (Colorado, U.S.A.)	940	24.8	24.1	8.00	1.93
Lat. 48½° N., Long. 17° W. (E. basin of N. Atlantic)	4,670[1]	0.116[2]	25.4	2.28	0.58
Lat. 46½° N., Long. 27½° W. (Central Valley of Mid-Atlantic Ridge)	4,109[1]	0.87[2]	315	2.07	6.52

1. Depth of water.
2. Measured temperature difference over depth range of 4.6 m for 1st and 2.8 m for 2nd station.

carried by thermal conduction, the amount emerging per unit area is equal to the product of the temperature gradient and the thermal conductivity. To indicate the orders of magnitude of the quantities in non-volcanic areas a few typical examples are given in Table 1. Table 2 gives typical conductivities of some types of rock: considerable variations are found depending on the composition and water content, particularly for sedimentary rocks.

TABLE 2. Thermal conductivity of rocks, 10^{-3} cal/cm s °C

Granite	6-9	
Dolerite	7-8	
Gneiss	5-9	normal to foliation
Gneiss	6-11	parallel to foliation
Quartzite	7-19	
Limestone	4-7	
Dolomite	9-14	
Sandstone	4-11	
Shale	3-6	
Rock Salt	13-17	
Wet ocean sediments	1.7-2.4	

For more extensive tables see Clark (1966).

In a place where there are no shallow sources the upward flow of heat should be the same at all depths accessible to the drill: the temperature gradient will therefore vary with the conductivity, being less in well conducting sections than in poorly conducting ones. It is not easy to find bores holes or mines in which reliable measurements can be made. Time must be allowed for the heat produced by drilling to dissipate itself; gas, oil, or water must not flow into or out of the hole; and cooling by mine ventilation must be avoided. Moreover in many places the emerging heat is not all transported by conduction: some may be carried by circulating water. A striking example is shown in Figure 3.

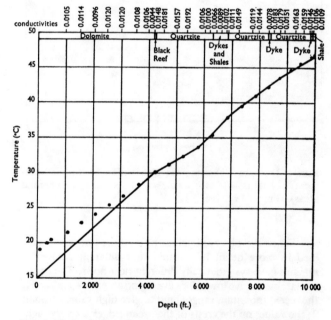

FIG. 3. The variation of temperature with depth in the Gerhardminnebron bore in the Transvaal. The dots show the observed temperatures and the full line the variation expected for a constant heat flow fitting the observations in the lower part of the hole. In the dolomite that occupies the upper 4,000 ft, the observed gradient is much less than expected: this is due to the free circulation of water in the fissures which carries part of the heat. (From: *Proc. R. Soc.*, ser. A, vol. 173, p. 489, 1939.)

At sea there is less possibility of the circulation of water in the sediment, and meaningful measurements can be made over a much shorter range of depths. Probes, 2 to 5 m in length, are forced into the bottom and by this means a large number of measurements can be obtained relatively quickly.

There are at present a few hundred places on land and a few thousand at sea where both temperature gradient and conductivity have been satisfactorily measured. The distribution of the values has been studied by Lee and Uyeda (Lee, 1965); the results are shown in Figure 4. The majority of the values give a more or less Gaussian distribution with a maximum at about 1.1 µcal/cm²s and with most of the measurements between 0 and 2.5 µcal/cm²s. On the high heat flow side of the peak there is a long tail extending beyond 8 µcal/cm²s. The distributions for the individual oceans and continents, taken separately, are very similar. The mean for the whole earth is 1.4 µcal/cm²s.

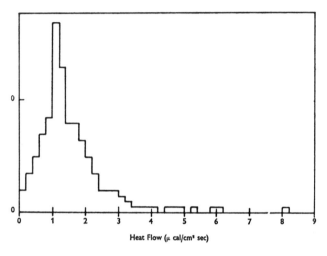

Fig. 4. Histogram showing the distribution of the measurements of heat flow for the whole world (excluding geothermal areas). (From: Lee, 1965.)

When the observed values of the heat flow are examined in more detail it is found that different geological regimes have systematically different heat flows. The continental shields give low values averaging about 0.9 µcal/cm²s; the recent mountain ranges tend to give high values around 2; the values on the crests of the ocean ridges are very high, often between 3 and 8; high values are also found in the inland seas behind island arcs, such as the Japan Sea. Very low values, around 0.6, are frequent in the oceans a few hundred kilometres on either side of the ocean ridges. Very high values are, naturally, found in volcanic and hydrothermal areas of the continents.

The source of the heat

A flow of 1.2 µcal/cm²s is not a large flux of heat by the standards of the kitchen or the factory. In a geological setting, however, it is impressive owing to the long period over which it is maintained. The earth is 4,500 million years old and of this the last 3,500 million years are represented by rocks near the present surface. In 1,000 million years a heat flow of 1.2 µcal/cm²s implies a total heat flow of

4×10^{10} cal/cm². If this were produced by the combustion of coal it would require the consumption of about 4 tons under each square centimetre of the earth's surface—that is the complete burning of a layer 20 km thick. Clearly the heat cannot be produced by burning coal or, indeed, by any chemical reaction near the surface, and clearly also some very powerful source of heat is needed to produce the observed flow over geological time-spans.

Some of the heat may be the original heat with which the earth was endowed when it was formed. Since the initial temperature is unknown there is some doubt about how large a part this may be. Lord Kelvin showed that if heat is brought to the surface by conduction in a solid earth, then, even if it was initially at the melting point, after 100 million years the original heat would be only a small fraction of that now observed. If heat is brought up by motions in the mantle, Kelvin's argument would need modification; it seems unlikely, however, that original heat contributes any large part of the observed flow. It is possible to suggest a number of other sources of heat within the earth, the outstanding one being the radioactivity of rocks.

All rocks contain small amounts of radioactive elements of which the only ones giving important amounts of heat are uranium, thorium and potassium and their products. Uranium has two long lived isotopes; U^{238} decays to lead (Pb^{206}) through a long series of intermediaries, and U^{235} decays similarly to Pb^{207}. Thorium has a single long lived isotope, Th^{232}, which decays, again through many intermediate stages, to Pb^{208}. The rare isotope of potassium, K^{40}, decays by two routes—one to Ca^{40} and the other to A^{40}. The half lives and heat productions are summarised in Table 3. The amounts of heat generated in different types of rocks are given in Table 4.

TABLE 3. Heat production by long-lived radioactive isotopes and their products

Isotope	Half life 10^9 yr	Proportion of isotope %	Heat generation[1] cal/g yr
U^{238}	4.50	99.27	0.70 ⎫
U^{235}	0.71	0.72	0.03 ⎬ 0.73
Th^{232}	13.9	100	0.20
K^{40}	1.31	0.012	27×10^{-6}

1. Per gram of the chemical element including all isotopes.

From the figures in Table 4 it is clear that there is no difficulty in accounting for the observed heat flow. In fact, since a thickness of 14 km of granite would produce a flow of 1 µcal/cm²s it is necessary to assume that the rocks below the crust are much less radioactive than crustal rocks; otherwise the heat flow would be much greater than is observed. Such an assumption is quite reasonable since the more basic rocks are much less radioactive than the granites.

TABLE 4. Typical heat production in rocks

Rock type	Concentration			Heat production (μ cal/g yr)			
	U ppm	Th ppm	K %	U	Th	K	Total
Granite	4.7	20	3.4	3.4	4.0	0.9	8.3
Basalt	0.6	2.7	0.8	0.44	0.54	0.23	1.21
Peridotite	0.016	0.004?	0.0012	0.012	0.001	0.0003	0.013

There is some direct evidence that the heat flow through the continental shield areas is, in large part, due to radioactivity in the crust. It is found that the heat flow is greater in areas of high radioactivity than it is in areas where the rocks are less radioactive (Roy *et al.*, 1968). Some results of this kind are shown in Figure 5; by extrapolating the observed values back to zero radioactivity, the heat flow for a non-radioactive crust can be estimated—i.e. the amount of heat coming from the mantle. It seems that on the continents this varies from one region to another in the range one third to two thirds. In the oceans there is only about 5 km of rather feebly radioactive basalt above the Moho and most of the heat must come from the mantle.

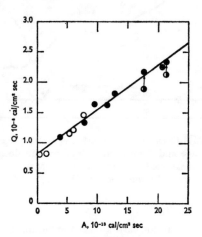

FIG. 5. Dependence of heat flow, 'Q', in Eastern and Central U.S.A. on the radioactivity, 'A', of the granitic basement rocks. (From: Roy *et al.*, 1968.)

Knowing the heat flow at the surface it is possible to make a rough estimate of temperatures all through the crust and in the upper part of the mantle, the main uncertainty being the distribution of radioactivity in depth. If, in a continental area, the heat flow is 1.2 μcal/cm²s, and if half of this comes from a crust 35 km thick with a conductivity of 0.005 cal/cm s °C then the temperature at the Moho would be 630 °C above that at the surface. Obviously

there is great uncertainty, but something in the range 500 to 700 °C seems reasonable. If the material below the Moho were static and the heat were transported by conduction, the gradient just below the Moho would be 12 °C/km and a temperature of about 1,400 °C would be reached at a depth of 100 km. The melting point increases with depth, but it seems likely that the temperature increases more rapidly and that the material of the mantle is near its melting point at depths of about 100 km. At greater depths the increase of melting point may exceed the rise in temperature and the material is probably in less danger of melting. The approach to melting at depths around 100 km is perhaps the cause of a slight decrease which is found in the velocity of elastic waves in this region.

Volcanoes, ridges and island arcs

The picture of an earth with a slightly radioactive crust and a still less radioactive mantle gives temperatures greatly below the melting point all through the crust. It is therefore an earth normally without molten rock near the surface and therefore without volcanoes or igneous activity in the crust. In a sense this is a good first approximation. Active volcanoes are exceptional features that occur in special places; they are not usual features of the landscape.

There are, essentially, two series of volcanoes: those around the island arcs and those on the mid-ocean ridges. To understand why this should be so is one of the central problems of geology. It is easy to see why volcanoes do not occur everywhere, but why should they occur on these two very different kinds of lines? Why also should the lavas of the two types be different? To understand these matters it is necessary to digress a little to explain the nature of mid-ocean ridges and island arcs and the current theories of their origin.

The mid-ocean ridges are shown in Figure 6; they are a great series of submarine mountain ranges which run around the world and are by far the longest chain of mountains on earth. The mid-Atlantic ridge is typical; some cross sections are shown in Figure 7. The ridge is completely different from a continental mountain range. Mountains such as the Alps are carved from great piles of compressed and distorted rocks, largely sediments; the mid-ocean

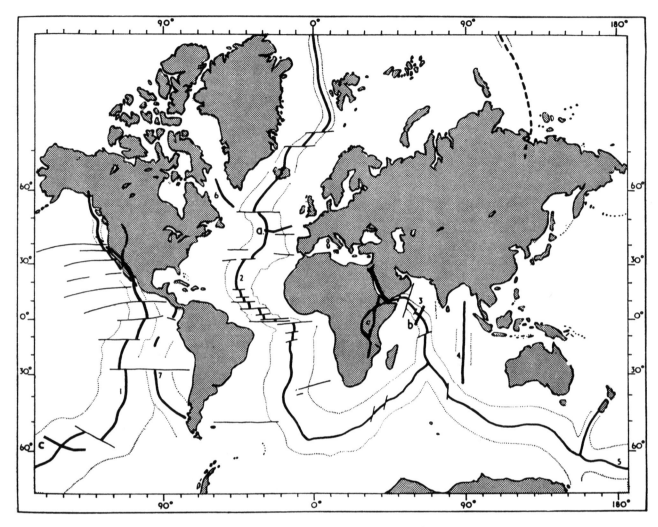

FIG. 6. The Mid-Ocean ridges. The heavy line represents the crest of the ridge and the central valley: the thin lines across the ridge are transverse faults. (From: D. H. Matthews, *International Dictionary of Geophysics*, p. 981.)

ridges, on the other hand, are composed entirely of submarine volcanoes. Along the crest of the ridge runs a steep sided, crack-like valley which is the seat of numerous earthquakes and also of volcanic eruptions. A study of the earthquakes has shown that the valley is an opening crack which opens a little more at each earthquake. As the crack opens, lava flows out and forms new volcanoes. It is not possible to discuss all the evidence here but there is now no doubt that the floor of an ocean behaves as two rigid plates that move apart by continual splitting along the axis of the ridge; reviews by Bullard (1969) and Menard (1969) provide more detail. In many places the ridge consists of sections offset along transverse faults, as may be seen from Figure 6. The sections of these transverse faults that lie between the two ridge crests have earthquakes; those further out, beyond

the crests, do not. In the earthquakes the motion is parallel to the fault as would be expected if two plates are separating along the ridge axis (Figure 8).

Outpourings of lava may occur at any point along the central valley of the mid-ocean ridge. Most of these will be beneath the surface of the sea and will pass unnoticed, but some will form volcanic islands. Such islands are seen at intervals along the whole length of the ridge; sometimes they are on the ridge crest but more often they are on the transverse faults. In the Atlantic there are Jan Mayen in the far north; Iceland, one of the few places where the central valley can be seen on land; the Azores where the volcanoes are spread out east-west along a transverse fracture rather then along the ridge crest; Ascension Island; St. Helena (well clear of the crest); Tristan da Cunha and Gough

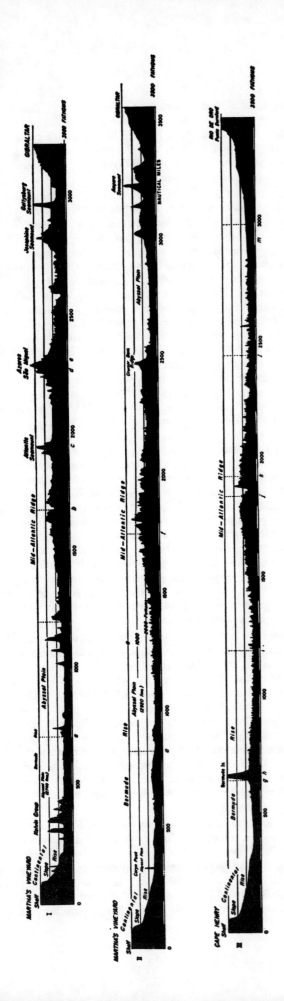

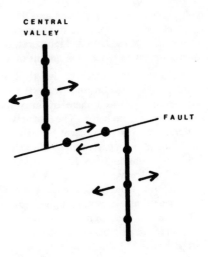

FIG. 7. Three sections across the North Atlantic Ocean, show-ing the mid-ocean ridge and its central valley. The vertical scale is exaggerated 40:1. The upper section passes through a transverse ridge from the Azores to Gibraltar: consequently the mid-ocean ridge appears less prominent than in the other sections. (From: D. C. Heezen and H. W. Menard, *The Sea*, vol. 3, p. 274.)

FIG. 8. Separation of plates along the central valley of a ridge. The motion must be parallel to the transverse fault. The earth-quakes (●) are confined to the central valley and the part of the fault between the two sections of valley.

Island. From the South Atlantic the ridge sweeps round to the South of Africa into the Indian Ocean. Here Marion and Prince Edward Islands are recently extinct volcanic cones. Further north Amsterdam and St. Paul Islands also lie on the ridge. The section of the ridge running north into the Gulf of Aden has no islands; it is continued in the Red Sea and is probably also connected with the eastern branch of the African Rift valley which shows at least one active volcano and many that are recently extinct. In the centre of the Indian Ocean the ridge forks: one branch goes around South Africa and one north as already described; the third goes south of Australia and New Zealand and enters the eastern Pacific. In the Pacific the ridge runs north and enters the Gulf of California. This section has no central valley and no active volcanoes; it has, however, a number of islands bearing recently extinct volcanoes, for example, those of Easter Island. The San Andreas fault runs from the northern end of the Gulf of California to Cape Mendecino where it runs out to sea. This fault is probably a transverse fault joining two sections of ridge, the southern piece in the Gulf of California and the more northerly piece running from near Cape Mendecino roughly parallel to the coast as far as the north end of Vancouver Island.

The course of the ridge has been described in some detail to show what part of the earth's active volcanoes lie on it; of areas that are actually being used for the production of steam or hot water we have Iceland and the Imperial valley of California. There are also a number of places where geothermal power could probably be developed if it were required; for example the Kenya Rift Valley, the Azores and perhaps the area of Abyssinia and Somaliland to the west of the Gulf of Aden. Unfortunately most of the ridge crest is beneath the sea and therefore inaccessible for geothermal development; the high heat flows measured in the central valley show that temperatures of several hundred degrees must be reached at depths of a few kilometres.

It has been said above that the ocean floors behave as rigid plates spreading out from the ridge axes and being continually renewed there by intrusion and extrusion of molten rock from below. The rate at which the plates spread differs from place to place, but is usually in the range of 1 to 10 cm/yr on each side of the axis. Such rates are, geologically speaking, very fast; spreading at 5 cm/yr on each side of a ridge would give a relative motion of 10,000 km in 100 million years. Since the earth is 4,500 million years old it is clear that plates must be destroyed as well as created. There is not much choice for the site of the destruction and all those who believe in moving plates believe that they are destroyed at the ocean trenches.

The Tonga trench is typical and has been particularly closely studied. The Tonga group of islands is a north-south chain of extinct volcanoes with a deep ocean trench running parallel to the islands and a little to the east of them. A plate is moving westward from the section of the ridge running north-south in the eastern Pacific. Spreading is fast, and in about 100 million years a piece of lava formed on the crest of the ridge would reach the trench. Here the moving plate dives down at about 45° and passes under the islands. The reality of this description is established by the earthquakes that occur within the plate as it moves and fractures. The earthquakes behind the Tonga trench occur at all depths down to 650 km. Earthquakes at such great depths occur only in places where there is reason to suppose that there are sinking plates; most are behind island arcs, such as those of Indonesia, Japan and the Aleutians, but they also occur beneath the Andes where there is an 'ocean trench' and a descending plate, but no island arc. The greatest depth of any earthquake is about 750 km; by the time the plate reaches this depth it will be near its melting point and unable to store the elastic energy needed to produce an earthquake. It is probable that the material of the plate finally becomes mixed again with the mantle material from which is was derived. The sediments on its surface may be partially scraped off and piled up just behind the trench and may contribute to the piles of distorted rocks found in the mountain ranges that often lie along the edges of continents. These mountains frequently contain volcanoes producing andesitic lavas, as in the Andes; and even where there is no great mountain range the volcanoes are often present, as in Indonesia. It is likely that the andesitic lavas are derived from the first products of melting of the sinking plates. Such melting would be expected at depths of about 150 km: with a plate dipping at 45° the lava would emerge at about this distance behind the trenches, assuming that it works its way vertically upwards from its source to the surface.

The areas of downgoing plates contain a large proportion of the geothermal areas of the world. Many of these are around the Pacific and include Indonesia, the Philippines, Japan, the Aleutians and the Andes. The New Zealand geothermal area is a southward extension of the Tonga trench and is probably associated with a sinking plate. Other areas where similar phenomena occur are the Caribbean, the Aegean and the Scotia Arc which runs between the southern tip of South America and Graham Land. There are some important geothermal areas whose positions in this scheme are not clear. The Mediterranean is an area in which very complicated events are in progress. Africa is moving northwards and colliding with southern Europe and the Middle East. In the collision the plates have become fragmented and directions of motion have changed. It is probable that the main, north-going African plate is going down and being consumed beneath the Cretan island arc, but it is not clear what is happening in Italy; it is possible that there is a small downgoing plate moving in from the west. Another area of doubt is in the mountain states of the U.S.A. The continent has moved westwards away from Europe and has probably overrun its own westward boundary trench and also part of a mid-ocean ridge. It is also possible that California is a continental fragment that had its origin somewhere to the west and has collided with the rest of the U.S.A. In view of these complexities, which are typical of continental geology, it is difficult to say what is the origin of the geothermal activity in northeast California and Nevada. It is clear that it is connected with the great outpourings of lava but it is not clear what is the origin and history of these.

There are many coastlines of the world where there are no earthquakes, no volcanoes, no downgoing plate and no geothermal areas. These coastlines, such as the eastern seaboard of the United States, the west coast of Europe and both the east and west coasts of Africa, are places where the ocean floor and the continent are parts of the same plate. The major plates are shown in Figure 9; they are delineated by the earthquakes of the mid-ocean ridges, which are associated with their generation, and those of the trenches which signal their destruction. There are in addition smaller plates in areas such as the Mediterranean which are imperfectly understood and cannot be shown in a small diagram.

In spite of the doubts about the meaning of geothermal phenomena in some areas, the suggestion that there are two main environments for igneous activity is a useful unifying concept in considering the phenomena. For any particular example it is useful to ask: 'Is this a place where the crust is splitting and new material is being added to a plate in the form of basaltic lavas and intrusions, or is it

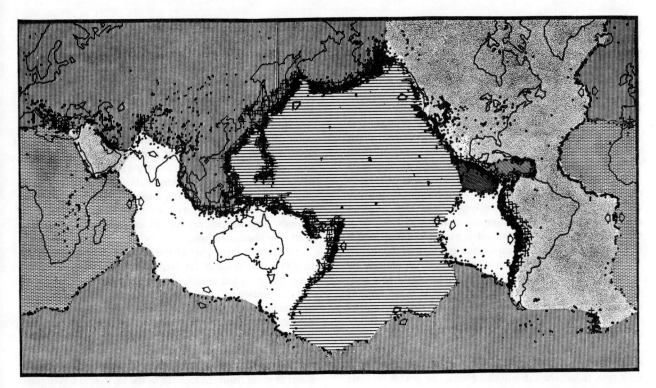

FIG. 9. The moving plates. The arrows show the direction of
motion. The plates are bounded by earthquake zones: shallow
earthquakes are shown by dots, and those deeper than 150 km
by crosses. Six large plates are shown—Europe/Asia, Africa,
Eastern Indian Ocean, Antarctic, North and South America,
and the Pacific. Some, but not all, of the small and medium
sized plates are also shown—South-east Pacific, Arabia,
Caribbean.

a place where a plate is being destroyed and melted and
andesitic lavas are working their way to the surface?'.

During the last year the ideas of 'plate tectonics' have
received important confirmation from drilling in the ocean
floor by the U.S. project J.O.I.D.E.S. (Joint Oceanographic
Institutions for Deep Earth Sampling). If our ideas are
correct, the youngest rocks should appear at the ridge crests
and in the central valley; some way on either side we
should expect recent sediments at the sea floor with older
sediments beneath. The age should increase with depth till
sediments are reached that were deposited when the rocks
at the drilling site were on the ridge axis. Below these sediments there should be lavas of the same age as these sediments. These predictions have been triumphantly verified
in the 130 holes so far drilled in both the Atlantic and the
Pacific oceans. The rates of spreading of the plates determined quite independently of the drilling give very good
estimates of the ages of the oldest sediments found in the
bore holes.

The mechanism of plate motion

Although we have now acquired a rather definite idea of
what is happening at the surface of the earth, we know very
little about the causes. We know how the plates are being
formed, are moving and are being destroyed, but what is
driving them? It seems likely that the earth is, in some sense,
a heat engine. Heat is continually being generated by the
decay of radioactive elements and this is likely to be the
ultimate source of the energy we see displayed in the motions at the surface.

If heat is generated in a layer of liquid the temperature
will rise and heat will flow out by thermal conduction. This
involves no motion of the material; but, if the rate of heat
generation and thus the temperature gradient exceeds a
limit known as the 'adiabatic gradient' the liquid becomes
unstable and motion starts. Hot liquid rises towards the
surface where it cools and sinks again. Such a motion is

called 'thermal convection'; once it starts, most of the heat is transported thereby and conduction is unimportant except in a layer near the surface. As has already been explained, it is likely that the mantle of the earth behaves as a liquid for long continued forces—that is, it has no long-term strength and no power to prevent convection once the adiabatic gradient is exceeded. The adiabatic gradient is '$g \alpha T/C$' where 'α' is the coefficient of expansion of the material, 'T' is the absolute temperature, 'g' the acceleration of gravity and 'C' the specific heat. All these quantities, except the coefficient of expansion, are quite accurately known; the coefficient of expansion of rock in the mantle will be less than it is at atmospheric pressure and, from extrapolation of experiments at high pressure and from theory, may be estimated at about $10^{-6}/°C$. With this value the adiabatic gradient is about 1 °C/km, which is certainly less than the temperature gradient in the mantle if it were static and the heat were brought up entirely by conduction. It is therefore not unreasonable to expect convective motions in the mantle. The form that these motions will take is difficult to predict: it depends strongly on the rate of change of viscosity with depth and with temperature and on the thickness of the layer taking part in the motion. The occurrence of long, narrow, linear features such as ridges and trenches suggests that the motion may be in long rolls rather than in isolated plumes or jets. A realistic theory would be quite complicated since it must allow for the thermal effects of the motions of the plates, for the effects of the cold sinking plates on the temperature distribution and for the differences in the distribution of radioactivity between oceans and continents. The heat engine may move the plates and may be the driving force of geological change, but we must also not ignore the effect of the moving plates on the heat engine. So far only rather simple models have been treated and much remains to be done before we have a realistic physical and mathematical theory of large scale geological processes.

The transfer of heat to the surface

As molten rock approaches the surface of the earth a very complicated series of processes will occur. The pressure will be reduced, gases and liquids held in solution will be released, the temperature will fall and crystallisation will begin. A lava is a complicated mixture of oxides and silicates and is capable of producing many different solid substances depending on the pressure, the temperature and the amount of water and other volatiles present. In general the first substances to crystallise will be those of highest melting point: these will be basic rocks without free quartz. The crystals will be denser than the liquid and will tend to settle to the bottom of the molten material; how far they are able to do so will depend on the size of the crystals, the viscosity of the melt and the geometry of the cavity in which it is contained. The composition of both the melt and of the separating crystals may also be affected by the

eating away and solution of material from the walls of the cavity. The details of all these processes form a large part of the subject matter of the science of petrology and cannot be discussed in any detail here. In general there will be a tendency for anything that will not form high melting compounds, or whose atoms will not easily fit as impurities into the common silicate minerals, to be left over as gases and liquids in the late stages of igneous activity. In particular, hot water and steam will be released and will carry away a great variety of other materials by 'steam distillation'. In this way many relatively rare elements may be concentrated. Deposits of elemental sulphur may be formed and its gaseous compounds, such as H_2S and SO_2 may emerge; even extremely rare elements with volatile compounds, such as mercury and silver, may occur as ores in workable concentrations.

In fact, a large part of the water and steam emerging in geothermal areas is not derived from deep seated molten rock but from the circulation of ground water. Lavas are full of joints and cracks and it is easy for rainwater to penetrate deeply and reach the hot lava; here it is itself heated and caused to rise, taking with it dissolved materials. When it reaches the surface it may emerge as steam or hot water or as a mixture of the two.

Some idea of the proportion of 'new' water—that is, water coming from the interior of the earth and not derived from rainwater—may be obtained from the study of the ratio of the isotopes of hydrogen in the water. Hydrogen has two stable isotopes; ordinary hydrogen, H, with atomic mass 1, and heavy hydrogen, D, (deuterium) with mass 2. The light isotope is much the more abundant. The ratio of the numbers of the two kinds of atoms in sea water is a little variable but is typically H/D = 6,400. Most of the deuterium will be combined in water molecules HDO; (since deuterium is so rare, there will be very few D_2O molecules). Owing to their greater mass the HDO molecules have a lower vapour pressure than H_2O molecules and, when water is evaporated from the sea, a higher proportion of the H_2O molecules will go into the vapour. It would therefore be expected that the water that falls as rain, which is derived from water vapour evaporated from the sea, would contain less deuterium than does sea water. This is found to be so: the ratio H/D in rain water, river water or tap water being typically 6,800. The ratio for water derived from geothermal steam is close to that for rainwater, and this suggests that most of the steam is derived from rainwater circulating through cracks in the rocks. The uppermost stage of a geothermal area is therefore a heat exchanger in which the heat of the rocks is transferred to circulating water. A review of these complicated matters will be found in Tongiorgi (1963).

Conclusion

Geothermal areas are of interest as sources of heat and power and they present important and challenging engineering problems which form an important part of the theme of this Review; but it is well to remember that the power extracted from a geothermal area is a manifestation of the workings of the great internal heat engine which drives the processes of geological change. It is not by chance that the geothermal areas lie where they do, but only in the last few years have we begun to understand the hidden workings of the engine and the reasons for what we see at the surface.

Bibliography

BELOUSSOV, V. V. 1970. Against the hypothesis of ocean floor spreading. *Tectonophysics* 9, 489. (In English.)

BULLARD, E. 1969. The origin of the oceans. *Sci. Amer.* 221, 66. (Also appears in the Scientific American publication 'The Oceans'.)

CLARK, S. P. 1966. Handbook of physical constants. *Mem. geol. Soc. Amer.*, no. 97, p. 459-482.

JEFFREYS, H. 1970. *The Earth* (5th ed.). Cambridge University Press. 525 p.

LEE, W. H. K. 1965. *Terrestrial heat flow*. Washington, American Geophysical Union. (Chapter 6 contains 'Review of heat flow data' by Lee and Uyeda.)

McKENZIE, D. P. 1968. The geophysical importance of high temperature creep. In: R. A. Phinney, ed., *The history of the Earth's crust*, p. 28-44. Princeton University Press.

MENARD, H. W. 1969. The deep ocean floor. *Sci. Amer.* 221, 127. (Also appears in the Scientific American publication 'The Oceans'.)

ROY, R. F.; BLACKWELL, D. D.; BIRCH, F. 1968. Heat generation of plutonic rocks and continental heat flow provinces. *Earth Planet. Sci. Lett.* 5, 1.

TARLING, D. H.; TARLING, M. P. 1971. *Continental drift*. London, Bell, 112 p.

TONGIORGI, E. 1963. *Nuclear geology in geothermal areas*. Pisa, Consiglio Nazionale delle Ricerche: Istituto Internazionale per le Ricerche Geotermiche. 284 p.

In addition to the above references the reader will find more general geological information in:

HOLMES, A. 1965. *Principles of physical geology*. London, Nelson. 1088 p.

For an account of the magnetic evidence for spreading of the ocean floor, see:

BULLARD, E. 1968. Reversals of the earth's magnetic field. *Phil. Trans. Roy. Soc.*, A, 263, 481.

II Exploration

The role of geology and hydrology in geothermal exploration

J. R. McNitt

Adviser in geothermal exploration
Resources and Transport Division
United Nations, New York (U.S.A.)

Introduction

This article describes the application of the sciences of geology and hydrology to the exploration of geothermal resources, and in particular, how geology is inter-related to the more specialised disciplines of geochemistry and geophysics. In contrast with geochemistry and geophysics, which use well defined techniques of data collection and rather standardised methods of interpretation, geology is more of a subjective discipline in which conclusions are based on a minimum of information that is often internally inconsistent. The experience and judgment of the geologist are therefore primary factors in determining the validity of his conclusions. Even geologic mapping, one of the few specific 'techniques' of the geologist, is far from standardised in its methods of implementation.

The objective of geothermal exploration is to find a reservoir of thermal fluid of sufficiently high temperature and permeability to produce economic quantities of heat and/or fluid to justify economic exploitation. The role of the geologist is to minimise exploration risk by ensuring that:

(a) the region selected for initial reconnaissance is likely to contain good prospect areas for detailed exploration;
(b) appropriate geologic, hydrogeologic, geochemical and geophysical data are collected and interpreted in order to select the most promising prospect areas and the best possible locations for exploratory drill sites;
(c) assuming a discovery is made, ensuring the collection of all appropriate well data so that reservoir parameters can be determined.

Thus the geologist must be a 'generalist', sufficiently familiar with a great variety of specialised exploration, drilling and well-testing techniques to ensure their proper coordination in the overall exploration effort. It is his task, in collaboration with the geophysicist and geochemist, to choose the specific exploration techniques which he believes will yield the best results under the given geologic conditions, and to resolve the inevitable conflicting interpretations resulting from the application of these various techniques. The bases on which geological decisions are made through the successive phases of an exploration project will be the principal theme of this article.

Selection of a region for reconnaissance

In selecting a region for preliminary reconnaissance, the best and most obvious criterion is the presence of thermal manifestations. In countries which have long established geological surveys, the locations and temperatures of such manifestations are often well documented, and an appropriate selection of a region for preliminary reconnaissance can be based on these records. Elsewhere, however, selection must be based on whatever is known about the geologic environment.

In order to assess the potential of various geologic environments for the production of geothermal resources, it is necessary to examine in some detail the environments of known geothermal systems. The general relationship of such systems to young orogenic zones, and in particular to recent volcanism within these zones, is well known. When examined in detail, however, this relationship is found to be far from simple. Some orogenic zones, particularly the island arcs, are intensely volcanic and contain a high density of high temperature manifestations. Other young orogenic zones, on the other hand, contain few volcanic centres and have a low density of low temperature springs. Examples of this type of orogen are the Alps, Taurus, Outer Carpathians, Southern Atlas, and the Eastern half of the North American Cordillera. To confuse matters further, there are regions having moderate to high temperature springs with no associated young volcanism, such as Central Anatolia and the Basin and Range Province of Western North America. Furthermore, there are young volcanic zones, such as the Cascade Range and the Hawaian Islands, which have very few hot springs. Needless to say, the general relationship between orogenic zones, volcanic centres, and thermal systems will require a great deal of investigation before it is thoroughly understood. However, new relevant information has been brought to light during the last decade.

Of the 43 geothermal fields proved by drilling, 27 (63 %) are associated with Quaternary volcanic centres. Of these, 11 are in structures related to volcanic processes, such as calderas and fractures peripheral to volcanic domes (e.g. Matsukawa and Otake, Japan; Matsao, Taiwan; Ahuachapan, El Salvador; and the Monte Amiata fields, Italy). The remaining 16 fields associated with Quaternary volcanic centres occur in fault block structures related to tectonic rather than volcanic processes (e.g. Wairakei and Broadlands, New Zealand; the Geysers and Salton Sea, U.S.A.; and Cerro Prieto, Mexico). The reservoir temperatures of the first group average about 220 °C and of the second group about 250 °C. These temperatures are usually found at depths of 500 to 1,000 m.

Ten of the 43 fields proved by drilling (23 %) have been found in areas unassociated with Quaternary volcanism. Of these, 5 are in 'hinterland'[1] fault block structures (e.g. Larderello, Italy; and Kizildere, Turkey); 3 in hinterland basins (e.g. the Hungarian Basin; and the inter-mountain basin of Georgia, U.S.S.R.); and 2 in rift zones (in Iceland). In the depth range of 500 to 1,000 m, the reservoir temperatures of the first group average 190 °C, the second 85 °C and the third 120 °C.

The remaining 6 of the 43 fields are located outside tectonic zones in 'foreland'[1] and platform areas, such as the Pre-Caucasus foreland and the West Siberia platform of the U.S.S.R. Unlike the fields discussed above, reservoirs in foreland and platform areas are not hyperthermal, but rather their temperatures reflect the world average gradient of approximately 30 °C/km of depth. From depths of about 2,000 m, however, reservoirs found in this environment are capable of producing considerable quantities of 60 to 100 °C water suitable for large scale industrial use or domestic heating.

The following table summarises this information and shows probable average temperatures to be expected in geothermal systems at 500 to 1,000 m depth in various geologic environments.

No absolute economic limit can be placed on the minimum usable temperature for a geothermal fluid, because this will depend on the use to which it is put and will vary from place to place according to the competitive cost of other local sources of energy. At the present level of technology, however, a minimum temperature of about 180 °C is required for power generation, although lower temperatures may be used for many industrial processes or for domestic heating. As power is the primary application of geothermal resources, the geology of only the first three environments listed in the above table will here be emphasised.

Because of the importance of fault block terrains in both volcanic and non-volcanic geothermal regions, their geology, and particularly their relation to the adjacent and generally non-geothermal tectonic zones, will be described.

Geologic environment	Average reservoir temperature to be expected, °C
1. Fault block terrains associated with Quaternary volcanism	250
2. Volcanic structures associated with Quaternary volcanism	220
3. Fault block terrains in Cenozoic hinterland regions, without Quaternary volcanism	190
4. Cenozoic Rift zones, without Quaternary volcanism	120
5. Sedimentary basins in Cenozoic hinterlands, without Quaternary volcanism	85
6. Foreland and platform regions, without Quaternary volcanism	30

Most tectonic zones have a distinct structural asymmetry imposed by the horizontal component of the stresses causing the deformation. The direction of the horizontal component of movement is reflected in the geometry of the deformed rocks, the asymmetry being most commonly seen in the preferred direction of overturning of folds and of thrusting. The region towards which the folds are inclined and the thrust directed is called the *foreland*, and the region away from which these structural elements are directed is called the *hinterland*. Hinterland regions are generally characterised by high heat flow, numerous hot springs and, in some cases, by widespread volcanic activity: it is in these regions, rather than the folded and thrusted zones, or the foreland regions, that high temperature geothermal fields are generally found.

In contrast to the folded and thrusted zones, hinterland regions are characterised by normal faulting which gives rise to fault block terrains. Geothermal systems appear to be genetically related to the normal fault zones in two ways:

(a) Normal faulting indicates tectonic extension of the crust which in turn may promote the intrusion of magma bodies close to the surface through some mechanism such as the release of confining pressure on rock already close to its melting point, and/or the provision of steeply dipping and deep penetrating fault zones up which the magma can rise;

(b) The faults also provide steeply dipping permeable channels through which surface water can circulate to considerable depth, become heated, and return rapidly to the surface.

Normal faulting does not give rise to a random series of horsts[2] and grabens[2], as commonly implied in general

1. The meaning of this term will be discussed later.
2. See article by G. Facca, Figure 6, page 68,

descriptions of these regions. The characteristic structural form resulting from the faulting is a series of tilted fault blocks. Belts of tilted fault blocks complement the asymmetry of adjacent folded and thrusted zones, where present. In the vertical plane, the asymmetry is expressed by the blocks being tilted in one dominant direction corresponding to the direction of tectonic movement in the adjacent folded and thrusted zone. In the horizontal plane, the asymmetry is expressed by arcing of the fault traces, the convex side of the arcs being dominantly oriented towards the direction of tectonic movement in the adjacent folded and thrusted zone.

Individual faults can be traced for several tens of kilometres over which distance their strike may swing through 10 to 30 degrees of arc. The concave side of the arc is consistently on the down-thrown side of the fault.

The throw of the fault, at least as measured on the surface, complements the degree of tilting of the associated fault block, that is, Quaternary strata may be offset only a few tens of metres and tilted only a few degrees; while Mesozoic formations may have stratigraphic offsets of several thousands of metres and tilts of 40 to 60 degrees. The width of individual fault blocks, measured perpendicular to strike, ranges from about 2 km to several tens of kilometres. The width of the fault belts may be a few tens of kilometres, as in the African Rift zones, or many hundreds of kilometres as in the Basin and Range Province of the North American Cordillera.

The convex side of the fault belt, i.e. the side towards which the blocks are dominantly tilted, is here termed the frontal border of the belt, on which the belt can be terminated by:

(a) a folded and thrusted zone, with the direction of axial plane tilting and fault thrusting corresponding to the direction of arcing and fault block tilting in the adjacent normal faulted zone. Examples of such terminations exist in various parts of the North American Cordillera from Western Canada southward through Idaho to Wyoming;

(b) a monocline that dips into a major structural depression, the monocline having the same direction of dip as the adjacent tilted fault blocks. Examples include El Salvador, where the monocline dips southward beneath the Pacific Ocean and appears to form the north flank of the Central American Trench; and northern California where the monocline forming the east side of the North Coast Ranges dips eastward into the Central Valley;

(c) a gently tilted plateau, again tilted in the direction of arcing and tilting in the adjacent fault belt. This type of termination is characteristic of the East African Rift zones, but is also found in North America (The Colorado Plateau) and in New Zealand (The Kaingaroa Plateau).

It is the author's opinion that although thermal manifestations and evidence of recent volcanism may be found throughout the faulted belt, the most likely place to find the largest and hottest thermal systems is along its frontal border. This is because the whole tectonic zone migrates with time from the hinterland towards the foreland region so that the frontal border of the normal fault zone is also its youngest part. A good example of this relationship is in North Island, New Zealand, where the largest thermal systems are located in an arcuate zone adjacent to the Kaingaroa Plateau, which borders the fault belt on the east.

Only those orogenic systems expressing unilateral asymmetry have here been described. Some orogenic zones have a distinct bilateral asymmetry, with the normal fault belt forming the central axial zone and the folded and thrusted zones flanking the axial zone on either side. The horizontal component of tectonic movement, as reflected by plateau tilts, monoclinal dips, inclined fold axes and thrust faults, is directed away from the central normal faulted zone on both its flanks. The axial region, or hinterland, of these bilateral zones is exceptionally wide and gives rise to vast regions suitable for geothermal prospecting. An excellent example of such a zone is the Alpine orogenic belt extending from Slovakia, through Hungary, western Roumania, eastern Yugoslavia and southern Bulgaria to Macedonia. From the Greek coast it crosses the Aegean Sea to the western coast of Turkey, and from there it crosses central Anatolia to western Iran. This whole region appears to be characterised by normal faulting, high heat flow, thermal springs and young volcanism. Its extension to the east of Iran, although not well documented, is highly probable.

Selection of prospect areas

Having chosen a promising region for prospecting, it is then necessary to select specific prospect areas for detailed exploration. Again, as in choosing the region itself, the best guides are thermal manifestations. If the locations of these are not well documented, they can be found in either of two ways, depending on the time and financial resources available. Where the exploration budget is small, thermal areas can be located by prospecting teams whose chief activity will be to interrogate the local population. Hot springs are such extraordinary and easily recognizable phenomena that, except for the most remote and uninhabited areas, their existence is generally known to the local people. This method is time-consuming and often frustrating, particularly after having spent many hours in search of what may eventually prove a minor seepage barely above ambient temperature; but it can be an effective and practical technique.

If sufficient funds are available, infra-red imagery can be used to locate thermal manifestations, and provided that interpretation of the resulting data offers no major problem, this method greatly shortens the time required for the inventory. Infra-red imagery has been used as a reconnaissance tool by the United Nations over an area of 15,000 sq. miles of the Afar in Ethiopia, but the results of this survey have not yet been published.

Having located the thermal manifestations, the following information should be collected:

(a) type; i.e. fumarole, steaming ground, spring, well, seepage, CO_2 vent, etc.;
(b) temperature;
(c) flow rate;
(d) local geologic control;
(e) chemistry.

The objective of collecting these data is to have a basis for comparing the relative merits of the various thermal areas under consideration and to choose the most promising for detailed investigation. In assigning exploration priorities it is most inadvisable to use only one criterion. All available information must be collected and interpreted in the context of what is known from rapid surface reconnaissance of local geologic and hydrologic conditions.

Surface manifestations may reflect conditions at depth directly, or very indirectly, depending on the extent to which the thermal system is masked by overlying non-thermal groundwater horizons. For example, it is tempting to use surface heat discharge as a first approximation of a system's size, and therefore its capacity to produce usable energy. This parameter, however, can be misleading. Such major thermal systems as the Geysers and Salton Sea fields in the U.S.A., the Larderello fields in Italy, and the Matsukawa field in Japan have very meagre surface heat discharge compared with the amount of energy released by drilling. Indeed, in many cases drilling has resulted not only in a greatly increased energy discharge, but has also proved that productive ground may extend several kilometres beyond a surface manifestation of only a few tens of metres diameter. In exceptional cases, the field has been found to extend 10 km or more beyond a small surface expression.

The capacity and spread of dry steam fields (vapour dominated systems) are particularly difficult to estimate from surface indications because the very existence of such systems requires a lithologic or hydrologic 'capping' which allows pressure, and therefore temperature, to build up in the trapped vapour. This same 'capping', however, prevents the escape of excessive discharges that could indicate the size of the system.

In contrast with vapour dominated systems, hydrostatic pressure is available in hot water systems and, consequently, high temperatures can be attained without the necessity of a tight 'cap'. If the top of such a system can reach the surface relatively unaffected by near-surface lithologic or hydrologic barriers, its surface heat discharge can be taken as an order of magnitude indication of the amount of energy that could be produced by drilling; and the extent in area of the manifestations, which would consist mainly of boiling springs, would be a fair indication of the area of productive ground. The Wairakei and Yellowstone thermal areas appear to be of this type. These systems, however, are in some ways unique. Many hot water systems are prevented from discharging directly to the surface by a variety of geologic and hydrologic circumstances.

Impermeable lithologic horizons can confine the top of a thermal system and deflect the up-flowing water many kilometres laterally. The closest point at which water discharges from the Ahuachapan thermal system, for example, is 7 km away from what is believed to be the centre of the up-flow. The water is marked at the surface only by steam-heated shallow groundwater. Loss of heat by boiling, and perhaps dilution with cold groundwater, lowers the temperature of the thermal fluid from 230 °C near the centre of the system to 70 °C at the closest discharge point.

A type of lithologic barrier quite commonly found over the top of thermal systems consists of fine lake or stream sediments deposited in the fault angle formed between tilted fault blocks. Some, or perhaps most, of the thermal fluid rising along the fault can be deflected by the impermeable sediments and travel up-dip beneath them to the opposite side of the lake basin or alluvial valley. At the Kizildere field in Turkey, thermal fluid is deflected about 5 km from the centre of up-flow by this type of geologic condition. Where springs occur at the contact of alluvium with bedrock, rather than on fault or fracture zones, deflection of this kind should be suspected.

Lithologic barriers are not the only features to prevent thermal systems from discharging directly to the surface. Local hydrologic conditions can also be a controlling factor. If, for example, the top of the thermal system is intersected by a large confined cold water aquifer, the heat will be swept down-gradient; and either it will be dispersed so that no evidence of the system ever appears at the surface, or the aquifer fluid may discharge at the surface as large-volume warm springs many kilometres distant from the actual source of heat. The correct interpretation of these large-discharge warm springs is difficult because they can be caused either by a remote, but localised, high temperature source, or be due to only a high conductive heat gradient extending over a large region. If the warm springs are related to the second mechanism, search for a high temperature thermal system could be fruitless.

Another type of hydrologic condition which greatly modifies the surface expression of a thermal system is where the top of the system intersects an unconfined, but quite deep groundwater body. If the groundwater is stagnant, or moving slowly, and the input of heat from the thermal system is sufficient to cause the groundwater to boil, surface manifestations will consist of patches of gently steaming ground unassociated with high pressure fumaroles at temperatures above the atmospheric boiling point. If the depth to the water table is great, of the order of 300 m or more, much of the rising steam will be condensed in the unsaturated zone, and consequently the amount of heat appearing at the surface will be considerably less than what is available at depth.

It will thus be appreciated that in order to use surface heat flow as a reliable criterion for selecting promising thermal systems for detailed exploration, it is first necessary correctly to interpret the local geologic and hydrologic conditions modifying the surface expression of the system.

Understanding these local conditions is no less important for a correct interpretation of geochemical criteria. As described elsewhere in this volume, modern research in geochemistry has made this one of the most effective reconnaissance tools for finding high temperature geothermal systems. Of particular importance is the use of dissolved silica and the ionic ratio of sodium to potassium in hot spring waters as indicators of reservoir temperature. But these criteria can be as misleading as surface heat flow if not evaluated in the light of local geologic and hydrologic environment.

For silica to be a valid indicator of reservoir temperature, the reservoir fluid must remain in the liquid phase as it rises to the surface. If a vapour dominated zone is formed above the liquid reservoir, the silica content of the steam-heated near-surface groundwater may indicate a much lower temperature than actually exists at depth. The presence in the spring water of dissolved constituents such as sodium and chloride is usually interpreted to mean that a vapour phase has not formed in the system because these elements are not mobile in a vapour phase at the temperatures usually encountered in hot spring systems. Silica temperatures of sodium chloride waters are therefore usually considered valid. A notable exception to this rule, however, is the Tiwi thermal system in Luzon, where a dry steam well has been drilled only a kilometre or so from a large sodium chloride spring. In this instance the surface springs appear to be steam-heated sea water, for which silica thermometry would of course not be applicable. A spring system now being investigated by the author in East Africa may be another example of steam heated chloride water. In this case, the springs rise around the shore of a highly saline lake. Reservoir temperatures inferred from the silica content of the springs are only 120 ºC, but these temperatures cannot reflect the true temperature of the system if the springs are actually saline lake water heated by a vapour phase rising from a deep thermal fluid reservoir. Additional geologic, hydrologic and geochemical data must be sought in order to resolve the problem.

The conclusion to be drawn from the examples here discussed is that no single criterion or 'rule' can be followed in selecting promising thermal areas for further prospecting. Each area must be considered individually, and valid conclusions can be reached only by evaluating all relevant data within the framework of the total geologic environment.

Selection of drill sites

Having selected a promising thermal area on the basis of surface manifestations, geology, hydrogeology and geochemistry, it is then necessary to undertake detailed surveys to determine the best location for exploratory drill sites. The number and depth of the exploratory holes will be determined by the results of these detailed surveys, the objective of which should be to locate the centre, or hottest part of the system and to determine the minimum depth to a permeable horizon or structure within the high temperature region.

First, a geologic map of the area should be made to a scale of about 1 : 25,000 and covering a sufficiently large area to show the relationship of regional structure to smaller secondary structures which might control the location of the surface manifestations. The overall objective of making the geologic map is to obtain a 3-dimensional 'picture' of the structure and stratigraphy of the area. Reasonable topographic control is essential as a basis for constructing geologic sections. In making the geologic map, particular attention must be given to determine the geologic control of the spring system; in searching for permeable horizons or structures which could be good production zones if intersected by drill-holes; and in identifying possible impermeable 'capping' horizons.

In wide alluviated areas where bedrock geology is not exposed, it may be necessary to use geophysical techniques, such as gravity, magnetic or seismic surveys, to determine the geologic structure controlling the thermal system. The Salton Sea and Cerro Prieto geothermal fields in southern California and northern Mexico are examples of areas where these geophysical methods have been quite useful in delineating geologic structure and in locating successful production wells. Most geothermal areas, however, occur in mountenous terrain where geologic structure and stratigraphy is well exposed, and geophysical surveys designed for the delineation of structure are unnecessary, at least in the early exploration stage. Later, after a discovery has been made, it may be desirable to make these surveys in order to investigate the deep structures underlying the reservoir and perhaps to identify buried magmatic heat sources.

Concurrently with geologic mapping, geochemical surveys should be undertaken on a more detailed scale than during the reconnaissance stage. The chemical surveys should be designed to investigate the various hydrologic conditions which could affect surface manifestations, as discussed above, and also to locate, if possible, the centre and hottest part of the deep-seated thermal system.

As many surface observations as possible should be made of the hydrogeology of the area during the course of geologic mapping so as to determine the relationship between cold, near-surface groundwater and the underlying thermal system. In some prospect areas, shallow water wells will already be available for gathering data on near-surface groundwater conditions. The hydrogeochemistry of surface waters and hot springs, however, will shed a great deal of light on these conditions, often without the need for shallow observation wells. Although shallow groundwater information can be very helpful, the ultimate questions which a hydrogeologic survey should be designed to answer are not concerned with shallow, near-surface conditions, but rather with the location of the re-charge area of the deep-seated thermal system, the storage capacity of the system, and the rate at which re-charge can balance withdrawal. On the other hand, because observations relating to these problems should extend over several hydrologic cycles and because tracer studies may require several years to complete, the survey should begin as soon as possible. Nevertheless, if obtaining these data requires drilling several deep observation wells, it does not seem economically prudent to begin a deep drilling programme designed to collect hydrologic data outside the prospect area until a geothermal discovery has been confirmed. A compromise is to begin at an early stage to assemble and assess all available surface and subsurface hydrologic data and to initiate periodic observations of springs and existing wells, if these data are not already being collected.

Although there may be sufficient geologic exposures and stratigraphic relief in the area to obviate the need for gravity, magnetic and seismic surveys, it is advisable to use more direct geophysical methods designed to delineate the extent in area and probable depth of the high temperature zone.

If the thermal area is large, and located in rugged poorly accessible terrain, it may be advisable to make an aerial infra-red imagery survey in order to ensure that all surface thermal manifestations have been located. The data from such a survey is also helpful for calculating the total heat flow from the area. It should be kept in mind, however, that the data resulting from this method, as well as the information from shallow resistivity observations and shallow temperature probes, can be greatly affected by near-surface geologic and hydrologic conditions, and may therefore have little bearing on the deep thermal system which is the real exploration target. For example, relatively high thermal gradients may be found by shallow temperature probes in impermeable strata overlying a thermal aquifer, but the same gradient could be measured over a shallow aquifer containing only moderately warm water as over a deeper aquifer containing much hotter water. Obviously, this method alone should not be used indiscriminately for choosing well sites, but should be used only in conjunction with another method which can determine the depth of the reservoir under investigation.

Of all geophysical methods used for thermal prospecting, the most successful appears to be deep resistivity. These surveys should be designed to obtain information from about 1,000 m, which is a reasonable depth from which to expect production. Even this method, however, cannot always be relied on because the low resistivity of thick shale sequences can mask the low resistivity in the search for chloride waters in volcanic areas of a hot saline reservoir; and large vapour dominated systems could go completely unrecognized.

The conclusion to be drawn from all this is the same as that on the selection of prospect areas: that is, never rely on one method alone, whether it be geologic, hydrogeologic, geochemical or geophysical. Individual surveys should be designed to answer specific questions, one survey being used to control and verify another. Only when all the surveys are completed, and as many of the internal ambiguities and inconsistencies resolved as possible, (it will not be possible to resolve them all), should well locations be selected.

Drilling and testing

It is not only the function of the geologist to select well sites, but also to assist the drilling engineer to programme the wells. Together they must decide on the depth and diameter of the well, the casing programme to be used, the frequency of collecting and the location of core samples, and the frequency and types of test measurements. Well before the drilling programme is started, a drilling engineer should join the exploration team in order to write the engineering specifications for the drilling equipment, casing and supplies and to ensure that well locations are properly prepared and that the necessary quantity of water for drilling is available at the well sites.

The number of wells required to explore any one prospect will depend on the number of drilling targets available and the size of these targets, as defined by the detailed surveys described earlier.

The depth of exploratory wells, as their location, must be determined on the basis of the preceeding surface surveys. The target depth may be established on the basis of geologic sections or from the recognition of significant horizons found from resistivity or seismic profiles. As geothermal reservoirs are often structurally complex, predictions of the depth to possible production zones can be highly inaccurate. It is therefore desirable to have drilling capacity available to reach depths of about 1,500 m, even though it may eventually prove necessary to drill to only 1,000 m or less. For this reason it is rarely desirable, in the author's opinion, to programme only for 'slim holes'— 4-in to 6-in diameter wells—because these wells can normally reach depths of only 700 to 800 m.

Adequate logging and testing of exploration wells is essential. The testing equipment should be at the site before drilling begins, because the manner in which the well should be completed can best be determined by the information provided by these tests.

Electric logging of geothermal wells is expensive, and therefore rarely done. The high cost is due to the fact that geothermal wells are usually drilled in regions where oil field logging services are not easily available and because electric logging of high temperature wells requires specialised equipment. The geologist, therefore, must make correlations and identify production intervals on the basis of lithologic, drilling time, and temperature logs, and (probably most important) the occurrence of lost circulation intervals.

As in all drilling, a lithologic log must be made from the drill cuttings, and rock type identifications should be checked at reasonable intervals by taking cores. The identification of lithologic boundaries can be facilitated in many cases by correlating the lithologic log with the drilling time log. In addition, every attempt should be made to correlate the lithologic log of the well with the stratigraphic section obtained from surface geologic mapping. Hydrothermal alteration of the material obtained from the well, however, often makes correlation very difficult, if not impossible. Indeed it is sometimes possible to correlate from well to well only on the basis of alteration mineralogy.

Spot downhole temperatures taken with maximum recording thermometers are not entirely adequate and should not replace temperature logs obtained with continuous recording instruments such as the Amerada type gauge. These gauges are not expensive, and do not require a field service team to operate, as do the electrical logging methods. Temperature logs should be run during drilling as often as prudent drilling practice allows, and daily for a period of several weeks after drilling is completed and before the well is allowed to flow. Retrograde temperature zones identified during drilling should not be cased off until it has been determined whether these are actually cold zones or whether they are hot zones which, because of good permeability, have accepted a large quantity of cool drilling fluid. Correlation of the temperature log with zones of lost circulation will help to resolve this question.

Permeable zones known to control thermal fluid should be tested quantitatively as they are encountered by running water injection tests at various flow rates and, at the same time, measuring down-hole pressure build-up with an Amerada gauge. This test will give an injectivity index which is a measure of formation permeability. Because geothermal wells tend to clean themselves during the first few days of production, injectivity indices are usually considerably lower than the comparable productivity index. Nevertheless, injectivity indices provide a means of quantitatively comparing one permeable zone with another in the same well. The same information can be obtained after the well is brought into production only by expensive and time-consuming tests involving the use of packers; or not at all, if the permeable zones have been cased off.

Production tests, which must be designed and supervised by a reservoir engineer, should be run as soon as the well is permitted to flow, because the results of a quantitative flow test are essential for deciding either to continue drilling additional wells or to abandon the area.

The decision to continue drilling additional exploration wells is an easy one if the first well has good production characteristics; if it is marginal, however, the decision may be extremely difficult. In such a case, a rational decision can be made only if it is known why the well is marginal. The three common reasons why a well either fails to produce, or produces only at a small flow rate, are low temperature, low permeability or a poor well completion.

If temperature is the problem and the chemistry of the thermal fluid from the well gives no indication of higher temperatures in the reservoir, serious consideration should be given to abandoning the area unless some subsurface information, hitherto unavailable, suggests that the well was not sited near the centre of the thermal system.

If downhole temperatures are found favourable, but the well fails to produce because of lack of permeability, several more holes should be drilled to search for a permeable production zone; and if one is found, to confirm its geometry and orientation. The majority of geothermal fields depend on fracture, rather than intergranular, permeability for production, and fracture permeability is notoriously erratic. This is another reason why initial exploration wells should be programmed as deep as available finances will allow. The deeper the well, the greater the chances of intersecting randomly distributed fracture zones.

If the problem is one of well completion, either the well should be reconditioned if possible, or another well should be drilled. Two common errors made in completing geothermal wells are 'mudding off' of the potential production zones, and casing off of upper production zones in an attempt to find better zones at depth.

In hot water fields, a fourth reason why wells may fail to produce is that the reservoir rest level is too deep easily to allow unloading, and consequent flashing, of the high temperature water. If the reservoir is large and the temperature of the field is high, it may even be technically and economically possible in such a situation to pump the fluid to the surface.

The exploration phase of the project does not end with proving the existence of a high temperature geothermal reservoir having sufficient permeability for production. The next, and probably the most difficult, question to answer is what is the capacity of the field. Unfortunately, the structural complexity and the importance of fracture permeability in most geothermal fields make capacity projections, based on only a few exploration wells, highly unreliable. For this reason, there is not a sharp division in geothermal projects, as in the oil industry, between exploration and development drilling. The problem of predicting production capacities of geothermal fields is circumvented by having sufficient exploration capital available to prove at the wellhead a sufficient amount of steam to operate a minimum size utilisation plant economically. For a power plant, this size could be in the range of 20 to 50 MW. Assuming an average production of 5 MW per well, there should be sufficient capital available, after a discovery is made, to drill five to ten additional offset wells. Experience has shown if 'production' rather than 'information' is the immediate goal, then these offset wells should be conservatively sited; that is, the offset should not be drilled more than 200 to 300 m from the discovery well.

Bibliography

Note. The reference 'Rome' denotes the United Nations Conference on New Sources of Energy, held in Rome in 1961 and published by the United Nations, New York, 1964 as Proceedings of that Conference.

The reference 'Pisa' denotes the United Nations Symposium on the Development and Utilisation of Geothermal Resources, held in Pisa in 1970.

ARANJO, E.; BUITRAGO, J.; CATALDI, R.; FERRARA, G.; PANICHI, C.; VILLEGAS, J. 1970. Preliminary study on the Ruiz geothermal project (Columbia). Pisa.

BARBIER, E.; BURGASSI, P.; CALAMAI, A.; CATALDI, R.; CERON, P. 1970. Relationships of the geothermal conditions to structural and geohydrological features in the Roccamonfina region (Northern Campania, Italy). Pisa.

BODVARSSON, G.; PALMASSON, G. 1964. Exploration of subsurface temperature in Iceland. Rome. G/24.

BOLDIZSAR, T. 1970. Geothermal energy production from porous sediments in Hungary. Pisa.

BONDARENKO, S.; MAVRITSKY, B.; POLUBOTKO, L. 1970. Methods for exploration of thermal waters and geological-commercial assessment of their deposits. Pisa.

BROWNE, P. R. L. 1970. Hydrothermal alteration as an aid in investigating geothermal fields. Pisa.

BUACHIDSE, I. M.; BUACHIDSE, G. I.; SHAORSHADSE, M. F. 1970. Thermal waters of Georgia. Pisa.

CATALDI, R. 1967. Remarks on the geothermal research in the region of Monte Amiata (Tuscany, Italy). *Bull. Volc.*, 30.

——; ROSSI, A.; SQUARCI, P.; STEFANI, G.; TAFFI, L.; STEFANI, G. C. 1970. Contribution to the knowledge of the Larderello geothermal region (Tuscany, Italy): Remarks on the Travale field. Pisa.

CHEN, C. 1970. Geology and geothermal power potential of the Tatun volcanic region. Pisa.

CORMY, G.; ARCHIMBAUD, J. D. d'. 1970. Les possibilités géothermiques de l'Algérie. Pisa.

——; ——; SURCIN, J. 1970. Prospection géothermique aux Antilles françaises, Guadeloupe et Martinique. Pisa.

DOWGIALLO, J. 1970. Occurrence and utilisation of thermal waters in Poland. Pisa.

FRANKO, O. 1970. The importance of information on hydrogeological structure and geothermal situation with respect to the prospection of the new sources of high enthalpy water in Slovakia. Pisa.

FROLOV, N.; VARTANYAN, G. 1970. Types of commercial deposits of thermal subterranean waters and some aspects relating to the assessment of their reserves. Pisa.

GRINDLEY, G. 1970. Sub-surface structures and relation to steam production, Broadlands geothermal field, New Zealand. Pisa.

HAYASHIDA, T.; EZIMA, Y. 1970. Development of Otake geothermal field. Pisa.

HEALY, J. 1970. Pre-investigation geological appraisal of geothermal fields. Pisa.

ISHIKAWA, T. 1970. Geothermal fields in Japan considered from the geological and petrological viewpoints. Pisa.

JONES, P. 1970. Geothermal resources of the Northern Gulf of Mexico basin. Pisa.

KHREBTOV, A.; TEN DAM, A. 1970. The Menderes massive geothermal province (Western Anatolia). Pisa.

KLIR, S. 1970. Geothermal areas of Czechoslovakia. Pisa.

KOUTAS, R.; GORDIYENKO, V. 1970. Thermal fields of Eastern Carpathians. Pisa.

MAKARENKO, V. F.; MAVRITSKY, B.; LOKSHIN, B.; KONONOV, V. 1970. Geothermal resources of the U.S.S.R. and prospects of their practical use. Pisa.

McNITT, J. R. 1970. The geologic environment of geothermal fields as a guide to exploration. Rapporteur's report, Section III. Pisa.

——. 1968. Exploration and development of geothermal power in California. *Spec. Rep. Calif. Div. min. Geol.*, 75.

MERCADO, S. 1969. Cerro Prieto geothermal field, Baja California. *Trans. Amer. geophys. Un.*, vol. 50, no. 2.

MONGELLI, F.; RICHETTI, G. 1970. Heat flow along the Candelaro fault—Gargano headland. Pisa.

MORI, Y. 1970. Recent plans of the geothermal exploitation. Pisa.

NAKAMURA, H.; SUMI, K.; KATAGIRI, K.; IWATA, T. 1970. Geological environment of Matsukawa geothermal area, Japan. Pisa.

PENTA, F.; BARTOLUCCI, G. 1962. Sullo stato delle 'ricerche' e dell'utilisazione industriale (termoelettrica) del vapore acqueo sotterraneo nei vari paesi del mondo. *R. C. Accad. Lincei*, Ser. 8. 32.

RADJA, V. 1970. Geothermal energy prospects in South Sulawesi, Indonesia. Pisa.

RAGNARS, K.; KRISTJAN, S.; SIGUNDUR, B.; EINARSSON, S. 1970. Development of the Namafjall area, Northern Iceland. Pisa.

ROSS, S. 1970. Geothermal potential of Idaho. Pisa.

SUKHAREV, G.; VLASOVA, S.; TARANUKHA, Y. 1970. The utilisation of thermal waters of the developed oil deposits of the Caucasus. Pisa.

SUMMERS, W. 1970. Geothermal prospects in New Mexico. Pisa.

TAMRAZYAN, G. 1970. Continental drift and thermal fields. Pisa.

TEN DAM, A.; ERENTOZ, C. 1970. Kizildere geothermal field. Pisa.

VAKIN, E.; POLAK, B.; SUGROBOV, V.; ERLIKH, E.; BELOUSOV, V.; PILIPENKO, G. 1970. Recent hydrothermal system of Kamchatka. Pisa.

VALLE, R. G.; FRIEDMAN, J. D.; GAWARECKI, S. J.; BANWELL, J. C. 1970. Photogeologic and thermal infra-red reconnaissance surveys of the Los Negritos-Ixtlan de los Hervores geothermal area, Michoacan, Mexico. Pisa.

WHITE, D. E. 1965. Geothermal Energy. *Circ. U.S. geol. Surv.* 519.

——. 1970. Geochemistry applied to the discovery, evaluation and exploitation of geothermal energy resources. Pisa.

WUNDERLICH, H. 1970. Geothermal resources and actual orogenic activity. Pisa.

YAMASAKI, I.; MATSUMOTO, Y.; HAYASHI, M. 1970. Geology and hydrothermal alteration of Otake geothermal area—Kujyu Volcano Group, Kyushu, Japan. Pisa.

ZEN, M. T.; RADJA, V. T. 1970. Result of the preliminary geological investigation of natural steam fields in Indonesia. Pisa.

Geophysical methods in geothermal exploration

C. J. Banwell
Consulting Geophysicist
(New Zealand)

Introduction

Geophysical prospecting may be defined as the art of detecting and interpreting anomalies in the local pattern of certain physical quantities, as measured by suitable sensing equipment and techniques. It is only to be expected that the presence of a geothermal field will affect or distort some of the local physical quantities, and geophysical aids have in fact proved to be of considerable value in the detection and interpretation of geothermal fields.

Geophysical work, however, should not be regarded in total isolation from other subjects. It must proceed in close coordination with geology, hydrology and geochemistry, so that physical measurements may constantly be interpreted and checked. For example, geochemistry can be used as a convenient *physical* instrument, like a sophisticated thermometer or steam detector, valuable both for planning and interpreting geophysical programmes. The location of promising drilling sites is an act of complex detection that must eventually be checked by actual drilling; but the intelligent weaving together of *all* survey data, geophysical and other, can often permit the drill holes to be sited to the best advantage. No single method of survey, be it geophysical, geochemical or geological, can be expected to yield a unique and unambiguous result, and the overall picture of a geothermal field and reservoir is built up by a continuous process of cooperative data synthesis and cross-checks. It will thus be necessary in the course of this article sometimes to stray beyond the subject of 'geophysics' in the strictly limited sense.

Preliminary evidence of a geothermal field

A geothermal reservoir may be defined as a zone in the earth's crust where the temperature, fluid pressure and permeability are sufficient to permit efficient exploitation. Furthermore, the size of the reservoir must be great enough to maintain the desired energy output for a long enough period, and the depth must be shallow enough to keep drilling costs within reasonable economic bounds. In terms of figures, this means that the mean reservoir temperature should generally be of the order of 200 °C or higher, the depth not more than about 2 km, and the volume at least 2 km^3. The fluid filling the pore spaces in the reservoir, if steam, would then have a pressure equal to or less than that of saturated vapour at the reservoir temperature (e.g. some 225 psia for 200 °C); or, if hot water, a base pressure of the order of the cold water hydrostatic—say, about 3,000 psia. In practice, a reservoir at the maximum depth and minimum temperature, as quoted above, would seldom be economic to develop: all reservoirs now being exploited on a significant scale lie at depths less than half the maximum and have temperatures appreciably above 200 °C.

A reservoir of even marginal size, temperature and depth would produce a steady state thermal anomaly at the surface, which could readily be measured by present techniques. It can therefore be said, at least in principle, that any reservoir of economic significance can be detected by surface exploration involving no more than shallow drilling and temperature measurements. In practice, heat transfer effects by fluid convection often modify the conductive temperature distribution to an important degree. Movement of cold surface water through permeable formations near the surface can sometimes distort, or almost obliterate, the conductive anomaly, so that gradient holes may have to be deepened sufficiently to allow the necessary temperature measurements to be made below the level of the disturbance. However, the great majority of hydrothermal systems known and explored up to the present are associated with extensive deep circulatory systems which result in discharges of hot water and steam at the surface, accompanied by strong surface temperature anomalies. They are thus readily detected by casual observation; and so many of these zones of surface manifestations are known in various parts of the world that it seems very probable that geothermal exploration, for some time to come, will be mainly concerned with the investigation of reservoirs whose presence has been betrayed by visible activity. This, however, does not necessarily mean that important thermal reservoirs may not exist without any significant surface evidence, or that they should not be sought by more refined methods of exploration.

World distribution of hydrothermal systems

Attempts have been made by investigators (Banwell, 1967) to define those regions of the earth that offer reasonable promise of geothermal exploitation. Areas believed to have sufficient potential to warrant power development consist of island arcs in the circum-Pacific belt, continental margins bordering the Pacific, parts of Southern Europe and Asia Minor, and transitional rock suites in South-west Italy, Sicily, Iceland, Jan Mayen, Spitzbergen and Central Africa. These areas are of course not necessarily exhaustive, but they contain 'prima facie' evidence of economic geothermal potential.

Most of these fields are associated in some way with vulcanism, although the connection is not always very close, as can be seen from the following list of fields already explored:

Field	Fluid	Source rocks
Larderello	Steam	Fractured limestones; dolomite.
New Zealand ⎫ Japan ⎬	Hot water	Acid volcanics.
The Geysers, California	Steam	Fractured greywacke.
Cerro Prieto, Mexico ⎫ Niland, California ⎬	Hot water	River delta sediments.
Pathé, Mexico	Hot water	Fractured middle tertiary volcanics.
Iceland	Hot water	Fractured cavernous basaltic lavas.
Northern Taiwan	Hot water	Acid volcanics and some sedimentaries.

There is no clear correlation either with particular rock types or with geological period, and it seems possible that geothermal fields are better sought on some non-geological basis. Possibly the common factor, if one indeed exists, lies in processes in the upper mantle or crust which in turn lead to areas of high tectonic activity, where magma bodies are injected into the crust and where continual crustal fracturing provides paths by which surface water can penetrate freely to heating zones, and also provides permeable reservoirs in which large quantities of heat and hot fluid (steam or water) can be stored. Fracturing may also result in the escape of some of the heat to the surface, thus providing direct indications of the presence of a hydrothermal system. It should however be mentioned that the deposition of minerals (especially quartz and calcite) from hot water systems will tend to seal the escape paths some distance below the surface, thus increasing the retaining capacity of the reservoir. Storage in *steam* systems, however, may sometimes be dependent on the presence of a pre-existing impermeable capping formation of suitable thickness: this possibly explains why hot water reservoirs of adequate size seem to be more common than steam systems.

Heat flow studies made by other investigators are also of great interest in that they indicate certain other regions where geothermal activity is apparent, though not necessarily of sufficient intensity to justify power generation. Lee and Uyeda (Lee, 1965) have demarcated certain areas of the globe where heat flow is abnormally high. These areas show no evident correlation with the distribution of geothermal areas mentioned in the early part of this section of the article. The most conspicuous zone of maximum heat flow is located over the eastern central part of the African continent. The maximum value is about 1.5 times the global average, and the maximum includes the zone of Ethiopian vulcanism and the Gulf of Aden, where high heat flows have been found on the sea floor. The thermal areas of Italy and Turkey lie near its northern margin, and an area of somewhat lower values extends into Hungary, where significant heat flow anomalies have also been observed. At the present time, little geothermal exploration data from Africa are available, and it would be premature to say whether this major anomaly has any significance of its own for power production.

Preliminary data required

For the initial planning of geothermal exploration before the execution of any true geophysical work, and for the preliminary selection of localised fields for detailed study, it is extremely useful to have a catalogue of basic physical and geochemical data from all the suspected fields. The compilation of a comprehensive catalogue of this sort could well be a major undertaking that would take too long to complete. But advance work in selected areas, where other considerations (e.g. economic, lack of alternative power sources, or special uses for geothermal heat) support early development, would greatly assist in the programming of further stages of the work. The preliminary data in question are those than can be collected fairly quickly, at low cost, and without the necessity for setting up an elaborate field establishment for each promising area. The information required is briefly listed below:

a) *Geographical*
   i) Location of field and apparent size of active area.
   ii) Topography. Access, roads, land ownership, special problems.
   iii) Hydrology. Drainage pattern, rivers, lakes, watersheds.

iv) Climate. Mean temperature and annual range, seasonal distribution of rainfall and annual total.

b) *Physical*

i) Location map of thermal features, hot springs, fumaroles, mud volcanoes, steaming ground, hot seepages into rivers and lakes.

ii) Temperatures and estimated discharges from all major features, overall heat and mass discharge.

iii) Suitability of area for shallow surveys (e.g. possible difficulties in penetrating hard surface rocks, disturbance of surface temperature pattern by surface run-off, etc.).

c) *Geochemical*

i) Sampling of principal thermal features (hot springs, geysers, fumaroles, etc.) for the determination of concentrations and ionic ratios of important dissolved and gaseous constituents; the selection of constituents to be determined being in accordance with the most recent recommendations and established methods of interpretation.

ii) Preparation of geochemical maps, showing the distribution of the more important element concentrations and ionic ratios over the whole field.

iii) Sampling of rivers and streams, where appropriate to determine the total rate of discharge of chemical constituents from the field.

The potential importance of the geochemical work as an indirect physical instrument can be very great. It has been shown that the concentrations of certain compounds and elements (e.g. SiO_2 and magnesium) and ionic ratios (e.g. Na/K) in the hot water can give useful preliminary indications of reservoir temperature. In addition, certain chemical equilibria (e.g. CO_2, H_2, CH_4, H_2O) in the gases discharged, and associated carbon isotope exchanges, are temperature-sensitive and can again be used for estimating underground temperatures. It may also be possible to use sulphur isotope equilibria in the same way. Thus several more or less independent ways of making preliminary estimates of reservoir temperature are available, and these may be used to check one another as well as the inferences from purely physical measurements. The presence or absence of certain other mineral constituents in the surface activity can also sometimes be used to discriminate between steam and hot water systems, and this kind of information is obviously very useful both for planning further survey work and for subsequent exploitation programmes.

For the planning and conduct of the next, more detailed, stages of geophysical survey it is of course necessary to give thought to the possible or probable ultimate purpose to which a useful geothermal field, if discovered, would be put. Points to be considered include whether there is a need for electric power only, or whether industrial or other uses could be developed—at least from low grade or waste heat; the minimum acceptable economic life of the field; whether the recovery of chemical components from the geothermal fluids could be economic; whether any special problems arise in the disposal of waste water; etc. Consideration of these problems could influence the choice between two or more promising fields on which exploratory efforts should first be concentrated.

First stage survey

GENERAL

Assuming that a review of the preliminary data has led to the conclusion that a potentially useful geothermal field has apparently been identified, the next object of exploration is to define its location, area and depth; also, if necessary, to improve the precision of some of the preliminary data, such as the heat and mass flows, and to carry out more detailed geochemical work.

MAPPING AND AERIAL SURVEYS

For the efficient performance of this stage of the work it is very desirable to have available good topographic maps, on a scale of 1 : 100,000 or larger and a set of recent stereoscopic aerial photographs on a scale of 1 : 10,000 for survey planning and positional control. Normal and infra-red colour photographs on the same scale can also be useful for outlining possible hot areas in the office before proceeding to the field. If the field is situated close to large bodies of water (lakes or the sea) or on the banks of a river, long wave infra-red scanner images of the area may be of value for mapping hot water seepages as well as the areas of intense ground surface activity. Recent experimental work shows that areas having a heat discharge of about 350 µcal/cm²s (roughly 230 times normal) and upwards can be identified with reasonable certainty. However, it is usually possible to map heat flow anomalies down to much lower levels by shallow surface thermometry, and a special aerial infra-red survey would be justified only in exceptional circumstances. Preferably, surveys of this kind, if they are to be made, are best undertaken at the same time as other aerial mapping and photographic projects.

At the outset of all field work it is important to set up a survey network and a system of benchmarks and permanent blocks at a sufficient number of points to allow all maps resulting from the surveys to be related to a common grid. Also it is very helpful for correlating the data collected to have all maps either on the same scale or, if this is inconvenient for some surveys, on the smallest possible number of different scales. If a precise regional map reference grid exists, maps should be related to this if practicable; otherwise a provisional local grid should be set up for later tying in to the principal network. Field parties should be provided with copies of printed or duplicated map blanks with the grid and showing the positions and levels of all reference blocks, trig points and major topographic features.

SURFACE TEMPERATURE AND SHALLOW RESISTIVITY SURVEYS

The object of this survey is to delineate the hot area and its margins more precisely, with a view to determining the conductive heat discharge, the location of the less conspicuous anomalies, and the correlation of the heat flow pattern with local topography, drainage, etc. The choice of method will depend on the type of surface cover predominant in the area. If it is mainly of loose pumice soil, ash or sediments, the 'one metre probe' method, using a perforating tool and bi-metallic thermometer, is both rapid and inexpensive. If the terrain is rocky or covered with hard sintered deposits it would be better to substitute shallow resistivity measurements in place of direct temperature readings. Since ground resistivity is influenced by temperature (Keller, 1970), the discovery of a resistivity anomaly can be indicative of a temperature anomaly. (Figures of only a few ohm-metres may be recorded in hot water reservoirs, as compared with several hundred ohm-metres or more in 'normal' areas.) The so-called 'electromagnetic gun', a portable inductive device, is convenient for carrying out a resistivity survey with a penetration of 25 to 30 m.

Subject to topography, it is generally convenient to carry out either type of survey (temperature or resistivity) with station spacings of 50 to 100 m. The larger spacing should still be close enough to avoid missing any features of importance, since a geothermal field of minimum economic size is unlikely to cover less than 2 or 3 km^2.

The form and position of an anomalous area should be considered in relation to possible direction of ground water movement, as indicated by the topography and general drainage direction through the area. If disturbances of this kind are large, the true position of the reservoir at depth may be considerably displaced upstream from the surface patterns, and possibly also from the area of surface activity.

ESTIMATION OF ENTHALPY

In most known hydrothermal systems, surface radiation and conduction account only for a negligible fraction of the total natural heat discharge from a thermal area. Practically all this discharge is brought up from below by convecting hot water or steam. Provided that the total mass and heat content of all fluids escaping from the area are accurately measured, and provided that proper correction is made for dilution by local cold surface water (and this can be done by chemical means—see Mahon, 1970), then it is possible to deduce the enthalpy of the fluid at depth from the ratio of heat flow to mass flow. It is important, for this purpose, to include the heat and mass losses from all warm springs and river seepages round the margins of the area. Some of these marginal seepages may be well disguised in rivers, lakes or general ground water movements across the area.

Physical enthalpy, measured in this way, may be compared with the value derived from geochemical estimates of reservoir temperature. If agreement is good, the result may be regarded as satisfactory; but if poor, a more careful check should be made of all possible routes of surface heat loss. An interesting example of this occured in the Broadlands field in New Zealand. Enthalpy calculated from observed heat and mass flows was at first much lower than that derived from drillhole temperature data. Further measurements revealed a large additional heat flow into the Waikato River from warm springs discharging very low chloride water made up of ground water heated by sub-surface steam. Because the chloride content of this water was much lower than that of the reservoir water, it had been missed. in the earlier chemical measurements made in the river upstream and downstream of the thermal area: only the physical measurements (precise temperature and flow measurements in the river) revealed the missing heat. With this correction the different estimates of enthalpy came into much closer agreement.

Second stage survey

GENERAL

This stage will probably proceed as a continuation of the first stage, or may even begin before the first stage is completed as soon as some preliminary data become available as to size and location of the shallow temperature anomaly.

DEEP ELECTRICAL EXPLORATION

This means resistivity measurements by electrical techniques involving current injection and voltage measurements by means of separate sets of electrodes in contact with the ground (as distinct from electromagnetic or induction methods which do not involve the drilling of any holes or even making electrical contact with the ground at any point). Most commonly used are the Schlumberger or Wenner electrode arrangements, which differ only in the spacing of the current and voltage electrodes in a single line.

The recommended method will generally be a series of Schlumberger traverses (van Nostrand and Cook, 1966) over the area of the surface thermal anomaly and as far outside as may be necessary to find apparently normal resistivity values. A 'traverse' is made by selecting a fixed electrode spacing and taking successive readings by shifting the whole set of electrodes along the survey line.

In addition to the traverses a number of soundings should be made at selected points to determine the vertical resistivity distribution and possible layering of different formations. A 'sounding' is made by keeping the centre of the electrode system at a fixed point and increasing the linear scale of the electrode pattern from low to high values.

'Traverses' are used for mapping the horizontal variations of the mean resistivity (taken over some chosen range of depths depending on the electrode spacing used) over the survey area. 'Soundings' give the variations of average resistivity as an increasing range of depths is included in the average. The graph of apparent resistivity against electrode spacing ('sounding curve') can be used to calculate the parameters (thickness and electrical resistivity) of the layers of a hypothetical underground model made up of a series of horizontal beds.

A supplementary geological survey may be of value at this stage if it can suggest the probable thickness and compositions of these layers. If clay, or clay-bearing formations, of significant thickness are present in the survey area, these may give low resistivity zones which are not associated with thermal anomalies. A geological study which can predict such formations and give their depth and thickness with reasonable precision can be of considerable value for interpretation.

The effective depth of penetration of resistivity traverses and soundings in low resistivity formations can sometimes be seriously limited by 'skin effect' if alternating current, or switched direct current of too short a period, is used. It is generally advisable to use direct current methods where possible. But if the equipment available permits only the use of alternating current, the penetration should be calculated from the frequency used and the measured resistivity, in order to avoid possible erroneous interpretations of the sounding profiles. It should also be noted that sounding profiles can be distorted if the soundings are carried out in a field of limited size, or too close to the boundary, the effect usually being to give an apparent increase of resisitivity at maximum spread.

Subject to these limitations the soundings should be carried out up to theoretical penetrations of about 3 km so as to obtain as complete a profile as possible. From the soundings it should be possible to deduce whether the reservoir is filled principally with steam or with hot water, thus checking earlier preliminary conclusions. In a hot water field the resistivity pattern takes the form of a central 'low', representing the porous reservoir rocks filled with hot chloride water, surrounded by a region of rapidly increasing resistivity outside the hot area. In a steam field, on the other hand, there would probably be a central resistivity 'high', because the reservoir rocks are now filled with steam which has low electrical conductivity, surrounded or overlain by lower resistivity zones containing steam-heated ground water and condensate. In actual fields, things are seldom quite so simple, and resistivity alone will seldom give a complete answer. As already stated, the presence of clay formations can behave deceptively like hot water reservoirs. Resistivity 'highs' within one area of low values can also sometimes be due to buried masses of impermeable rocks such as rhyolite domes, massive lava flows, etc.: these can sometimes be detected and mapped by seismic (refraction) surveying, by gravity or magnetic mapping, or from geological evidence. Thus evidence based on resistivity should always be cross checked against evidence based on other methods.

An apparent hot water reservoir can be readily mapped by means of traverses, using a current electrode spacing of about 500 m, giving an effective penetration of some 250 m. With a steam reservoir it is possible that overlying confining formations will have a lower resistivity than the reservoir rocks, which may therefore be identified by a resistivity rise in the soundings. Owing to the limitations mentioned above, however, checks by thermal gradient measurements are advisable.

Before leaving the subject of deep resistivity measurements, it may be mentioned that variants of the electrode arrangement (e.g. the so-called 'dipole-dipole system') can be used with good effect for mapping horizontal discontinuities (e.g. faults, boundaries of hydrothermal systems, geological contacts, etc.) and this can often serve for checking proposed geological models. Geological models can also sometimes be used for the preliminary interpretation of sounding curves.

Electromagnetic methods have also been proposed for deep resistivity surveys, though rather different approaches would be needed (Keller, 1970; Strangway, 1970). These may have useful application after further development, but not much use has yet been made of them for geothermal exploration. (See also Meidav, 1970.)

THERMAL GRADIENT DRILLING

Holes for thermal gradient measurement must be drilled deep enough to penetrate any surface formations liable to be disturbed by ground water movement, and they should extend far enough into the undisturbed zone below to give reliable gradients. Although it may sometimes be possible to estimate the thickness of the disturbed layer from geological or other evidence, it is very desirable to check whether a hole is deep enough by measuring temperatures at multiple points spaced about $2\frac{1}{2}$ m apart over the bottom 20 m of the hole. If the measured temperatures fall on a straight line when plotted against depth, the deduced gradient is probably reliable; otherwise, the hole must be deepened or the area avoided. In general, very little will be learned by drilling gradient holes too close to the areas of surface activity; the gradient will be subject to disturbance by upward movement of hot water or steam, and the holes will be liable to erupt.

If gradient holes are drilled on a regular grid pattern with a spacing of 1 km, a thermal anomaly of reasonable size (covering, say, 10 km²), could be mapped, inclusive of margin areas, with an array of 20 to 30 holes. A hole density of this order should suffice for all but the most detailed mapping, since experience has shown that lateral temperature variations below the surface convective zone are generally not very rapid. Additional holes could, however, be necessary in some fields if marked discontinuities in the temperature pattern indicate corresponding structural features, such as faults, contacts, etc., which might in some cases also be checked against geological data. As a general rule these gradient holes would be too shallow (30 to 40 m) to provide much new information about structure, but coring for geophysical or geological information could be profitable, at least on a trial basis, in a few holes.

For actual temperature measurements various instruments may be used. For temperatures much over 100 ºC a geothermograph or Amerada gauge is suitable. For lower temperatures it may be convenient to use thermo-couples, thermistors, platinum resistance thermometers, or even mercury maximum thermometers with proper precautions. Electrical instruments can be wired up differentially so as to measure the gradient directly, though this is seldom done in practice. Care must be taken, with electrical instruments, to ensure that the cable insulation is suited to the temperatures to which it is exposed. Nylon or pvc is suitable for moderate temperatures, while mineral insulation or such materials as 'Teflon' may be required for higher temperatures.

Third stage survey

SUNDRY GEOPHYSICAL AIDS

Little or nothing has hitherto been said of the following geophysical techniques which have been used or recommended from time to time for geothermal exploration:

> Dipole resistivity
> Magnetic
> Gravity
> Seismic (reflection or refraction)
> Microwave
> Radio-frequency interference
> Ground noise
> Micro-seismicity

This is because it will often be unnecessary to use any of these aids to bring the investigation to the point where a deep exploratory hole can be planned and sited. On past occasions there has perhaps been a tendency to undertake some of these specialised surveys solely in the belief that they may perhaps be useful in some way or another. In this way, considerable expenditure can be incurred of money, time and effort in producing maps which, though of potential *research* value, may do little or nothing to promote the search for an exploitable geothermal reservoir.

Nevertheless, in an actual survey problems could arise which some of these techniques could help to resolve; they could perhaps account for anomalies in the temperature or resistivity patterns. But the choice of technique, and the justification for using it at all, must arise in, and be defined by, the progress of the basic survey. Thus a seismic survey could be of possible value if it permitted any discontinuities or other features of the resistivity soundings to be identified in terms of other physical properties, or if it enabled their depths, extent and thickness to be checked. Seismic surveying could also have potential value in locating zones of low velocity, low wave frequency or high attenuation, which could be associated with a steam or hot water reservoir (Hayakawa, 1970). Similarly a ground noise survey, though its precise application is still under trial at present, might

be used to check the reservoir location suggested by other surveys; and a geological study would be of value if it could provide relevant advance information about the physical properties of the underground formations (porosity, permeability, density, seismic velocity, depth thickness and continuity).

ESTIMATION OF RESERVOIR CAPACITY

Completion of this stage of the exploration programme should enable a first estimate to be made of the power potential, using the family of curves published by Banwell (1964). Taking the area indicated by the deep resistivity and surface gradient measurements, and the reservoir temperature implied by the geochemical studies, it is possible to deduce from the curves the approximate field potential in megawatt-years. Thus, for example, a thermal anomaly with a horizontal area of 10 km^2 and an inferred reservoir temperature of 260 ºC would have an estimated potential of about 4,000 MW-years, or sufficient to maintain an output of 200 MW for 20 years. This estimate is believed to be conservative, as it assumes a hot water system with a boiling point/depth distribution and an energy yield of only 25% of the theoretical.

Fourth stage survey: deep exploratory drilling

The foregoing programmes should have outlined the surface projection of the presumed steam or hot water reservoir, its depth and probable temperature, and the sequence of formations to be traversed, together with some indication of their physical properties. The value of a deep hole is to test the proposed model by providing quantitative data, first from cores and later from temperature measurements, followed by production tests if the hole taps a productive zone. Projected drilling depth will depend on expected reservoir depth and thickness, but it is suggested that a minimum of 1 km should be planned for unless special circumstances should suggest otherwise.

If it is thought important that steam should be produced from the first hole, it will be desirable to try to locate it so as to intersect permeable formations at reservoir depth. A connection between surface fault traces and the existence of permeable zones has been claimed in some fields. It has also been suggested that the presence of both vertical and horizontal fissures, which could provide gas or steam paths, can be predicted by seismic records. The use of the surface fault criterion, taken alone, is apt to be severely limited by the fact that in most areas a very large number of fractures and fault traces can be mapped from aerial photographs and surface surveys, whereas very few, if any, of these may prove to have any immediate connection with thermal activity or reservoir permeability. There may perhaps be some general connection between surface faulting and reservoir permeability in that the presence of numerous young faults above the reservoir could indicate a good probability of drillholes intersecting permeable zones in the reservoir, but it would be unwise to depend on any closer relationship. In the absence of more definite information on the distribution of reservoir permeability, holes may be sited in relation to surface fault traces as well as by any other means; but the success or failure of the first few holes, however sited, to give satisfactory production, can give only very limited information about overall reservoir permeability.

Possibly, the only general criteria for the siting of deep exploration holes are as follows:

(i) The sites should be far enough away from areas of surface activity and hot ground to avoid the risk of blow-outs at shallow depth in the early stages of drilling and casing;

(ii) Subject to (i), most holes should be located well within the boundaries of the reservoir as indicated by deep resistivity and gradient surveying;

(iii) Despite (ii), some test holes should be drilled well out on the boundary of the apparent reservoir, to check the correctness of its limits and the form of the temperature/depth curve near the boundary;

(iv) Holes should be well distributed over the reservoir area, without too many in one line, so as to permit a three-dimensional 'model' of the reservoir to be prepared as early as possible.

The coring programme for these holes should be planned to provide enough samples to allow the relevant physical properties of each of the formations identified by the preceding surveys to be determined. In addition, samples are needed from zones characterised by their physical state —e.g. temperature, presence of steam or water in the pores, or cementation by mineral deposition or transformation. Petrological examination of core material can also show significant mineral changes due to hydrothermal alteration, and the nature and extent of these changes can give an indication of the temperature at which they took place.

The core data can now be used, in conjunction with the models derived from the previous surveys, to construct a series of sections showing the distribution of physical properties through the volume surveyed. These sections can then be used to interpret and refine the earlier survey data and possibly to suggest other surveys that could be profitable. It could be decided, for example, whether any features of the density distribution are important for the further evaluation of the field, and whether they will give large enough anomalies at the surface to justify a gravity survey. Core data can of course also be used to construct geological sections if required.

After completing each hole, temperature measurements should be made at regular intervals over the full depth until stable values are reached. Aquifer pressure measurements should also be made at several levels, assuming the hole is partly filled with water, and chemical samples taken from levels of interest. After this, the hole, if hot enough, should be allowed to discharge, the output should be measured, and the steam and water should be sampled for chemical analysis.

Data from the first few holes will have an increasing influence on the siting of those to follow. Also, if reservoir conditions prove favourable, an increasing number of these holes will be available for production, with the result that the diameter, depth and placing of many new holes will be decided by engineering and field layout considerations. However, data from some of these production holes will still be worth gathering to improve the picture of the field and of reservoir geometry and conditions; while the chemistry of the water and steam produced, and the changes in this chemistry that occur as exploitation proceeds, will give further important information about the reservoir, its relation to its surroundings, and its probable useful life.

Effects of exploitation on the field

As the exploitation of the field proceeds, the withdrawal of steam or water from the reservoir and the inflow of recharge fluid from outside will affect the hydraulic balance and pressure distribution, both in the reservoir and in the surrounding areas. Pressure changes can be monitored by regular measurements in the various prospecting and production holes, while the mass balance can be checked by surface gravity measurements. To do this satisfactorily, it is advisable to set up a network of permanent gravity stations over the area, preferably before production begins, and to carry out precision gravity surveys (to within about $\pm\,0.01$ milligal, or better) at intervals of approximately one year. Comparisons between the gravity values from these annual surveys will show within quite close limits whether the reservoir is being depleted by draw-off, and to what extent the fluid removed by exploitation is being replaced. The exploitation of a field can eventually result in measurable ground level changes.

Bibliography

BANWELL, C. J. 1961. Geothermal drill-holes: physical investigations. *Proc. U.N. Conf. on new sources of Energy. Rome, 1961*, vol. 2, p. 60-72 (E/CONF 35/G/53). New York, United Nations.

——. 1967. Geothermal power. *Impact Sci. Soc.*, Paris, vol. 3 XVII (1967) no. 2, p. 149-166, table 2.

DOBRIN, Milton B. 1960. *Introduction to geophysical prospecting*. New York, Toronto, London, McGraw Hill. i-ix + 446 p.

HAYAKAWA, H. 1970. The study of underground structure and geophysical state in geothermal areas by seismic exploration. *U.N. Symp. on Geothermal Energy. Pisa, 1970.*[1]

HUNT, T. M. 1970. Net mass loss from the Wairakei geothermal field, New Zealand. *U.N. Symp. on Geothermal Energy. Pisa, 1970.*[1]

KELLER, G. V. 1970. Induction methods in prospecting for hot water. *U.N. Symp. on Geothermal Energy. Pisa, 1970.*[1]

LEE, W. H. K. 1965. *Terrestial heat flow.* Washington, American Geophysical Union (Monograph no. 8). See map, fig. 44, by Lee and Uyeda.

MAHON, W. A. J. 1970. Chemistry in the exploration and exploitation of hydrothermal systems. *U.N. Symp. on Geothermal Energy. Pisa, 1970.*[1]

MEIDAV, T. 1970. Application of electrical resistivity and gravimetry in deep geothermal exploration. *U.N. Symp. on Geothermal Energy. Pisa, 1970.*[1]

NOSTRAND, R. G. van; COOK, K. L. 1966. Interpretation of resistivity data. *Prof. Pap. U.S. geol. Surv.*, 499. Washington, U.S. Government Printing Office. (Library of Congress catalog-card no. GS 65-322. For sale by Superintendent of Documents, U.S. Government Printing Office, Washington D.C. 20402.) i-xi + 310 p.

STRANGWAY, D. W. 1970. Geophysical exploration through geologic cover. *U.N. Symp. on Geothermal Energy. Pisa, 1970.*[1]

1. Published by the Istituto Internazionale per le Ricerche Geotermiche, Lungarno Pacinotti 55, Pisa.

Geochemical methods in geothermal exploration

Gudmundur E. Sigvaldason
**Science Institute,
University of Iceland, Reykjavik**

Introduction

In the past ten years geochemical methods have been applied increasingly to evaluate various physiochemical parameters which are important feasibility functions in the planning and operation of a geothermal development. Geochemical methods are now widely used in preliminary prospecting for potential geothermal exploitation, and chemical data on natural discharge from thermal areas serve as an important guide for decision making on subsurface exploration by drilling. As drilling proceeds, chemical analysis of deep thermal fluids provides information on flow patterns of water and assists in selecting improved drilling sites. During production, testing, and subsequent utilisation, chemistry provides an efficient and inexpensive tool to detect minor and major changes in the reservoir, both with regard to temperature and water level fluctuations.

The wide applicability of geochemistry in all stages of geothermal exploration is especially important because of the relatively low cost involved as compared to geophysical surveys and subsurface investigations by drilling. This is especially important during the prospecting stage. Geochemical studies of natural thermal indications in a virgin area can very often give a conclusive answer as to the feasibility of further investment in that particular area.

The usefulness of geochemistry in geothermal exploration was long obscured by a scholarly dispute about the origin of thermal water and the nature of its dissolved chemical elements. Most of the published studies centred on the question of magmatic or meteoric origin, and attempts were made to relate individual elements or components to some theoretical source. In the 1950's it was shown by isotope studies that most of the water in thermal areas anywhere in the world could be directly related to meteoric water. A small, but significant, amount of 'magmatic' or 'juvenile' water could, however, not be excluded and the dispute continues to this very day.

In the 1960's an entirely new approach was initiated putting the main emphasis on the fact that a thermal area represents a chemical system at high temperature. The main components in this system are a fluid phase, the water, and a complicated, heterogenous solid phase, the wall rock. No matter what the origin of the heat or the individual chemical elements in the system, reactions between the components would seek equilibrium at a relatively fast rate because of the high temperature. The chemistry of the thermal fluid could therefore not be interpreted in terms of a magmatic or other diffusely defined origin. The chemistry reflects rather a state of last equilibrium between the fluid and the solid phases of the system, and can be interpreted accordingly.

Types of thermal waters and gases

Very many chemical analyses of thermal waters and gases have been published over the last decades, especially from those areas where economic interest has been involved. Reviews of chemical composition of thermal water have been made by White (1957a and b), White, Hem, and Waring (1963), and Ellis and Mahon (1964). White (1957a) grouped hot spring water into a few chemical types; and a detailed discussion, also considering rarer types of subsurface waters, is given by White, Hem, and Waring (1963). Table 1 gives examples of the most important types of thermal water.

SODIUM CHLORIDE WATER (Table 1, nos. 1, 2, 3)

Sodium chloride water is the most common type in large underground reservoir systems. The pH is close to neutral at depth, but becomes slightly alkaline as the water comes to the surface and loses steam and CO_2. The commonest anion is chloride and the ratio chloride/sulphate is high. Sodium is the principal cation. A relatively wide concentration range is found within this group.

ACID SULPHATE CHLORIDE WATER (Table 2, nos. 4, 5)

Acid sulphate chloride water is a relatively rare type of thermal water. The acidity of the water is due to oxidation

TABLE 1

	1	2	3	4	5	6	7
SiO_2	501	640	456	412	322	109	60
Li	n.d.	14.2	n.d.	n.d.		n.d.	2.3
Na	250	1.320	5,025	609	75	2.0	129
K	25	225	905	51	11	3.0	69
Rb	n.d.	2.8	n.d.	n.d.		n.d.	
Cs	n.d.	2.5	n.d.	n.d.		n.d.	
Ca	0.9	17	354	14	263	2.2	272
Mg	0.0	0.03	23.4	4	73	0	68
F	9.5	8.3	1.5				2.4
Cl	127	2,260	8,730	878	1,490	15	170
Br	n.d.	6.0	n.d.				
I	n.d.	0.3	n.d.				
SO_4	108	36	28	262	3,730	758	501
As	n.d.	4.8	n.d.				
B	n.d.	28.8	131	4.4		6.9	4.3
NH_3	n.d.	0.15	n.d.	2		30	1.0
HCO_3	133	19	49				667
CO_3	70						
H_2S	0.2	—	—	—	216	—	2.6
pH	9.26	8.6	7.02	3.1	1.6	1.97	6.6
T °C	100		(220)	55	81	90	72

1. Great Geysir, Iceland (Sigvaldason, 1966).
2. Wairakei, New Zealand, Hole 44 (Ellis and Mahon, 1964).
3. Ahuachapan, El Salvador. Drill hole (Magnesium value is obtained by complexometric titration and probably too high).
4. Frying Pan Lake, Tarawera, New Zealand. (White, Hem, Waring, 1963.) Al: 4, Fe: 2.
5. Yang Ming Shan, North of Taipai, Taiwan. (Cited from White, Hem, Waring, 1963.) Al: 600, Fe: 95, PO_4: 2.9.
6. Norris Basin, Yellowstone Park, U.S.A. (Cited from White, Hem, Waring, 1963.) Al: 2.4, Fe: 0.8.
7. Mammoth, Yellowstone Park, U.S.A. (Cited from White, Hem, Waring, 1963.) Al: 0.2, Fe :0.06, As: 0.5, Sr: 0.4.

Note. 'n.d.' means 'not determined'.

of sulphide to bisulphate at depth. As the waters rise to the surface and cool down the pH shifts from neutral to acid because of change in the dissociation constant of bisulphate with temperature (Ellis and Wilson, 1961).

ACID SULPHATE WATER (Table 1, no. 6)

Acid sulphate water is common in fumarolic areas, where steam rising from a hot water reservoir condenses at the surface. H_2S contained in the steam is oxidized upon contact with air and also as a result of bacterial activity (Schoen and Ehrlich, 1968) to form H_2SO_4. The water has a very low chloride content and may contain large and varying amounts of cations derived by acid leaching of small rocks. The springs containing acid sulphate waters usually have little or no discharge, and often the water carries much suspended clay material.

CALCIUM BICARBONATE WATER (Table 1, no. 7)

Calcium bicarbonate water occurs commonly as low temperature travertine depositing springs. Water of this kind

has not been found at economically feasible temperatures, and the calcite precipitation would be a severe drawback in utilization.

Within the same geothermal system more than one type of thermal water can occur. An important aspect of hydrothermal chemistry is to find a logical relation between the different types of water within the thermal area, and especially to show how the chemistry of hot spring water can be used to predict the chemical composition of thermal waters at depth and the physical environment of the water in the geothermal reservoir.

The thermal gases form another set of chemical groups. The same type of thermal gas may appear with two or more distinct types of thermal waters. Three main types of thermal gases can be defined: (1) gases characterized by very high N_2 and little or no active gases; (2) gases with very high CO_2, but minimal H_2S and H_2; and (3) gases with high H_2 and H_2S, but CO_2 also as a major constituent. A continuous gradation between (2) and (3) exists and even to some extent between all groups.

Table 2 gives an example of chemical analyses of thermal gases from a few major thermal areas.

Table 2.

	1	2	3	4	5	6
CO_2	0.0	92.2	91.0	86.7	67.0	63.5
H_2S	0.0	4.2	2.6	4.1	7.3	1.69
H_2	0.01	1.8	2.0	2.6	23.7	14.67
CH_4	0.04	0.9	0.03	0.3	0.2	15.29
N_2	97.1					
Ar	2.07	0.3	4.43	6.3	1.4	3.53
O_2	0.0	—	0.0	0.0	0.0	
NH_3	—	0.6	—	—	—	1.28
H_3BO_3	—	0.05	—	—	—	0.14

1. Reykjavik, Iceland. Drillhole.
2. Wairakei, New Zealand. Drillhole (Ellis, 1967).
3. Hengill. Hverakjálki, Iceland. Steam vent.
4. Hengill, Hveragerdi, Iceland. Drillhole.
5. Hengill, Nesjavellir. Drillhole.
6. 'The Geysers', California, U.S.A. (White, Hem and Waring, 1963).

Origin of thermal waters and gases

One of the most important contributions to our present knowledge of geothermal systems was the recognition of the meteoric origin of thermal waters. Craig, Boata, and White (1956), and Craig (1963) studied the isotope ratios H/D and O^{16}/O^{18} in thermal waters from widely separated thermal areas and compared these with the isotopic composition of meteoric waters in the corresponding areas (Fig. 1). The points for meteoric water are distributed over a wide range on the curve according to climatic conditions and geographic position of the area. The H/D values for the thermal waters coincide in each case exactly with the H/D values for the meteoric waters, but the O^{16}/O^{18} values are offset to the right due to exchange of oxygen isotopes in the thermal water with oxygen of the wall rocks in water channels. These results went a long way to settle the dispute about the volcanic or 'juvenile' character of thermal waters, but the authors pointed out that a small amount (5-10%) of water other than meteoric in origin could not be excluded. This leaves open the possibility to interpret the origin of dissolved components in the thermal water as being of volcanic origin, injected into the main body of meteoric water as a high density steam carrying large amounts of dissolved solids.

In a series of papers Ellis (1968), Ellis and Mahon (1964, 1967), and Mahon (1967) have described hydrothermal experiments which show that a large part of the dissolved solids in thermal waters can be explained on the basis of reactions between water and wall rock of permeable channels in the thermal water reservoir. Two types of reaction can be defined: (1) reactions where a temperature-dependent solution equilibrium is established between water and minerals of the wall rock; and (2) leaching of elements not accomodated in the structure of stable mineral phases at the physiochemical conditions in the reservoir. From these simple rules it is possible to predict two major characteristic features which should be common to all deep circulating thermal waters: (1) all elements governed by solution equilibria should be in approximately equal amounts in all thermal waters, and the range in composi-

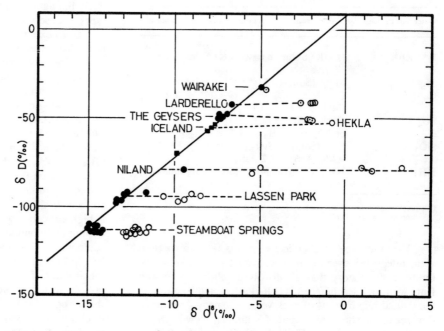

Fig. 1. Isotope variation in sodium chloride type thermal waters. (From: Craig, 1963.)

51

tion should be in accordance with variation in temperature; (2) all elements leached from wall rocks, and which do not precipitate in secondary mineral phases, should show large spread in values between individual thermal areas. This spread will naturally depend on such factors as availability of the element in rocks, and the ratio between rock volume and the volume of water which acts upon the rocks. A glance at Table 1 shows that the prediction holds very well. Comparing for example the values for silica and chloride it is evident that while chloride varies within wide limits the range of silica values lies between 100 ppm and 600 ppm SiO_2. The range in silica values spans the solubility of silica in the observed temperature range of thermal areas. The chloride values on the other hand exhibit an extremely wide range which, considering the relatively even distribution of chloride in the earth's crust (excluding the oceans and evaporate deposits), could be explained on the basis of rock/water ratios in the geothermal systems. Low chloride water would be derived from a permeable area with large throughflow of water, or an old thermal area where the water has been confined to stable channels over long periods of time. High chloride would be expected from an area with low rate of flow, where a large surface area of rock is exposed to a relatively small volume of water. A reservoir losing its water in the form of steam would also tend to build up a high concentration of chloride, since chloride is not transported in the steam phase.

Table 3 is a simplified summary of experimental data on the behaviour of individual elements under hydrothermal conditions.

TABLE 3

Water constituents	Temperature and pressure dependent equilibria
K^+, Rb^+	(Na/K) Solution $\rightleftharpoons$ (Na/K) Feldspar, mica /Rb /Rb
Mg^{++}	Mg^{++} (Solu) $\rightleftharpoons$ Mg^{++} (Clay minerals)
Ca^{++}	$CaCO_3 + CO_2 + H_2 \rightleftharpoons Ca^{++} + 2\ HCO_3$
F^-	$Ca^{++} + 2\ F^- \rightleftharpoons CaF_2$
SO_4^{-2}	$Ca^{++} + SO_4^{-2} \rightleftharpoons CaSO_4$ or $SO_4^{-2} + 4\ H_2 \rightleftharpoons S^{-2} + 4\ H_2O$
SiO_2	SiO_2 (Solu) $\rightleftharpoons$ SiO_2 (Quartz)
As	$As_2S_3 + S^{-2} \rightleftharpoons 2\ AsS_2$
NH_4^+	NH_4^+ (Solu) $\rightleftharpoons$ NH_4^+ (Secondary minerals)

Chemical equilibria which control some of the constituents of natural thermal waters (1966).

The discharge of chemical elements from a thermal area, integrated over its lifetime, would give quantitative limits to the theory that all dissolved solids are derived from wall rock leaching. Ellis (1966) attempted to estimate the annual requirement of rock volume to provide some of the ions in the chemical output of Wairakei, New Zealand. For the minor elements, Cl, B, F, Li, Cs, and As, 0.02 to 0.0001 km^3 of rhyolite were needed per year, and the heat

flow from the area could be maintained by the crystallization of 0.01 km^3/year of magma. Grindley (1964) assumes the age of the Wairakei area to be 500,000 years and the volume of rhyolite, which has been totally leached with regard to some of the above elements, would be 2,500-5,000 km^3 if the rate of the chemical flow has been the same over the entire lifetime of the system. Since the rock porosity is probably of the order of 10%, the total affected rock volume would have to be much greater. Craig (1962) suggested, on the basis of carbon isotope data, that water circulation times are between 30,000 and 300,000 years in Steamboat Springs, Nevada. Considering such a slow water movement the throughput of water during the lifetime of the geothermal system would be only a few successive fillings of the reservoir (1966).

Geochemical prospecting

The great majority of explored geothermal systems are associated with considerable surface activity. It is to be expected that in the near future economic interest will be focused on areas with noticeable surface indications and the following chapters will discuss mainly the geochemical relation between surface activity and reservoir. It should be realized, however, that completely hidden reservoirs undoubtedly exist.

Two principal types of surface activity are associated with a hydrothermal system: fumarolic, or steam fields, with insignificant discharge of water; and hot spring areas, often with considerable discharge of thermal water. Both types of activity sometimes occur together within a relatively limited space, but in other areas only one of the two types is observed.

Deep exploration of numerous thermal areas of both types has revealed the relationship between the different types of surface activity and the reservoir at depth. Numerous complicating factors can affect the general picture, especially with regard to chemistry of surface activity, but the following model is widely accepted:

A geothermal system consists of three principal constituents: (1) a heat source; (2) a heat transfer medium, the water which is contained in a porous rock formation; and (3) a cap rock, which serves as a confining lid to the system. Depending on the condition of the cap rock it may or may not allow some of the contained water to escape to the surface. If the hydrostatic pressure of the reservoir water is high enough it will flow out at the surface, adjusting by boiling and conductive heat loss to the new pressure environment. In this case the water appears on the surface as hot springs at boiling temperature or less, depending on the amount of heat loss and possible admixture with cold ground water. If the water table is at some depth below the topographic surface, boiling may occur along zones of weakness in the cap rock, resulting in steam escape into the formation, and possibly generation of fumaroles or steam outlets at the surface.

Both types of surface activity can therefore result from the same reservoir, depending on the general hydrology of the area and the surface topography.

Judging from this simplified model, discharging hot springs do provide samples of the deep reservoir fluid. This is only partially true, since the reservoir fluid may become mixed with ground water at shallow levels and always changes its chemistry by leaching and reaction with wall rock on the way to the surface. The main object of geothermal prospecting is to find clues in the chemistry of hot springs to the chemical and physical properties of the reservoir fluid.

Of primary interest is the relation between hot spring chemistry and reservoir temperature. Those elements in the thermal water which are governed by temperature dependent solutions equilibria can be used to estimate the subsurface temperature. These are principally silica, magnesium, and the ratio between sodium and potassium.

Figure 2 shows the solubility of quartz and amorphous silica in water. Silica in waters obtained at different temperatures in hydrothermal experiments with common rock types (Ellis and Mahon, 1964) are also reproduced in the diagram. Obviously it is not easy in any single case to select with certainty the appropriate curve, but the possibil-

ities can be considerably narrowed. Studies of hydrothermal alteration in explored geothermal areas have shown that quartz is always an important secondary mineral phase at relatively shallow depth (Steiner, 1963; Sigvaldason, 1963; Muffler and White, 1969). In most cases one would therefore compare the silica value of the spring water with the quartz solubility curve. This is done assuming that silica remains metastably in solution through the temperature gradient from the reservoir to the surface. This may actually be so if the speed and volume of water is considerable, but reliable estimates of temperature by the silica method cannot be expected from springs with low discharge. Any precipitation of silica on the way to the surface would result in low temperature values. Higher values, on the other hand, would result if additional silica were dissolved from the wall rock at depths above the zone of quartz precipitation. The silica value of the water might then approach the solubility for amorphous silica. The silica thermometer should therefore be used with caution when dealing with surface springs, especially when the spring has a low discharge or is a minor seepage.

The atomic ratio of sodium to potassium is another indicator of reservoir temperature. This ratio is governed by a complicated equilibrium involving alkalifeldspars and

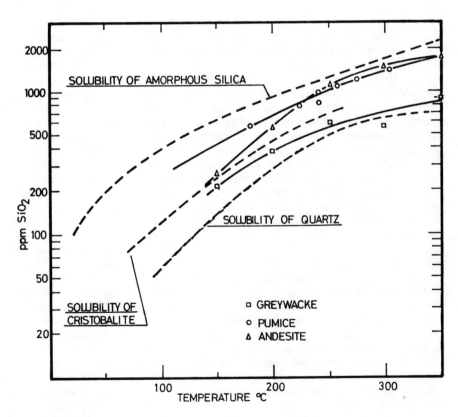

FIG. 2. Solubility of various modifications of silica and common rock types. (From: Ellis and Mahon, 1964.)

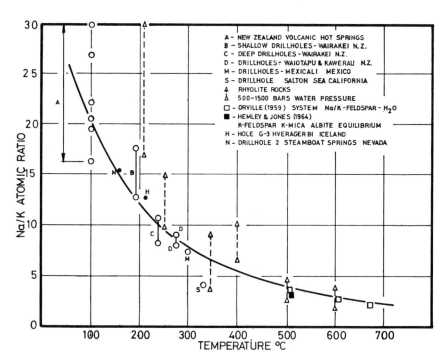

FIG. 3. Atomic ratio Na/K for thermal waters from various sources and experimental reaction solutions. (From Ellis and Mahon, 1967.)

K-mica (Ellis and Mahon, 1967). Figure 3 shows the Na/K-T °C diagram which in the higher temperature range is based on experimental data (Hemley and Jones, 1964; Orville, 1963) but in the lower temperature range the chemical data are based on drill hole discharges with known down-hole temperatures. Essentially the same cautionary remarks apply to the use of the Na/K atomic ratio for surface hot springs as for the silica equilibrium concentrations, for Na and K take longer to establish than the SiO_2 equilibrium, which would give a longer metastable persistence of the Na/K ratio. On the other hand potassium is preferentially absorbed on the surface of hydrothermal clay minerals such as montmorillonite. This mineral, and the closely related mixed layer structures montmorillonite-illite and chlorite-vermiculite, are common in the upper few hundred meters of geothermal areas and potassium uptake into these minerals may affect the Na/K ratio.

Solubility curves for magnesium in secondary minerals have not been determined, but hydrothermal experiments show that magnesium is preferentially incorporated in clay minerals which are stable at high temperature, such as chlorite. As a result, water which has been above 200 °C in contact with altered rocks is highly depleted in magnesium, the Mg concentration seldom exceeding 1 ppm. Magnesium does not therefore give a well defined temperature value, but its absence indicates reservoir temperatures which are economically feasible.

All of these temperature indicators, as well as other elements in the hot spring waters, may show more or less regular gradation within a group of hot springs. Chemical gradation can reveal important information on shallow subsurface flow patterns of thermal water and the location of zones of major upflow. Furthermore, they will assist in judging which particular hot spring within a given area comes closest to showing major factors in the chemistry of the reservoir fluid.

Silica and the alkali ions are probably the best indicators of near surface changes due to precipitation and ion exchange with wall rocks. Thermal water leaving a reservoir at 250 °C and appearing at the surface at 100 °C is highly supersaturated with silica, both because of convective heat loss and direct concentration and cooling by boiling off about 20% steam. Very often the water will approach the surface by a defined channel, or channels, that may become dispersed along horizontal near surface layers and flow out on the surface in numerous boiling springs. Silica precipitation will proceed proportionally to flow distance from the main upflow channels, and this will be reflected in the silica concentration of the springs. Figure 4a shows this clearly for the hot spring group around the Great Geysir in Iceland. Figure 4b shows the same pattern for the Geysir group based on the potassium values. The sodium and chloride concentration is constant in all the springs.

The application of chemical gradation to determine subsurface flow patterns can be extended to give a three-dimensional picture as soon as the thermal fluid can be sampled from various depths in producing drill holes. Ellis and Wilson (1960) located two major zones of upflow

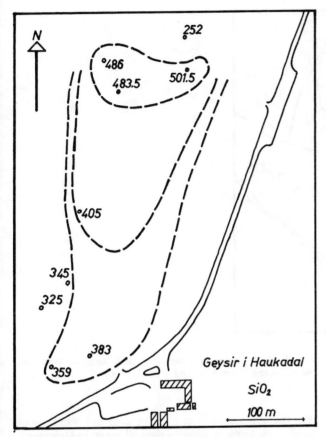

FIG. 4a. Variations in silica in surface hot springs due to precipitation in near surface layers.

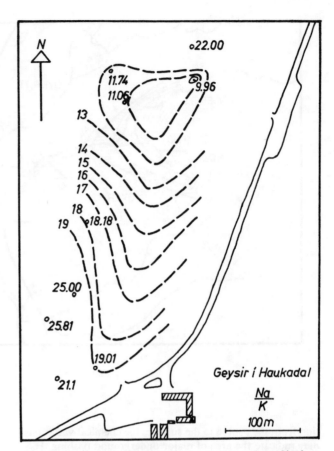

FIG. 4b. Variation in Na/K ratio due to absorption of 'K' on clays in near surface layers.

in the Wairakei field in New Zealand using the ratios Na/K, Na/Rb, and Li/Cs. Figure 5 shows the variation in the Na/K ratio on a map of the area where the results are projected in a horizontal plane. Under favourable conditions it may therefore be possible to locate upflow zones in a thermal area by chemical methods. Such information will help in locating sites for the first drill holes in a new area and can assist in selecting improved drill sites in areas which are being developed.

Fumarolic fields or steam fields with little or no discharge offer fewer possibilities for chemical prospecting than do the hot spring areas. The steam is often associated with some H_2S which tends to oxidize near the surface to sulphuric acid. The acid attacks the surrounding rocks and turns them into soft clays, usually containing kaolinite, opal and some sulphates (Sigvaldason, 1959; White and Robertson, 1962). The zone of acid alteration extends down to the ground water level where an abrupt change to alkaline condition occurs, resulting in totally different mineral assemblages. The condensation water which accumulates in the steam pits is diluted with meteoric water, depending on the amount of precipitation in the area (Árnason et al., 1969). The chemistry of the steamfield water does not there-

fore have a direct relation to an underground reservoir of thermal water, except with regard to volatile components. The water has a pH value ranging down to 0.5 or less, and due to its chemical agressiveness it often contains large amounts of various elements which for the most part are derived from the surrounding rocks.

The volatile components do not provide direct information on the reservoir temperature since little is known about the relative amounts of volatiles in relation to temperature. Empirical evidence does, however, indicate that the presence of hydrogen in the fumarole gases is associated with temperatures in excess of 200 °C.

In Table 2 are listed examples of the most common types of thermal gases. High nitrogen gases (> 90% N_2) are never associated with fumarolic activity, but gases consisting principally of CO_2 may be associated with both low temperature calcium bicarbonate springs and high temperature fields. The most common gas composition in fumarolic fields is CO_2, H_2S and H_2 in various proportions. A survey of gas chemistry in individual steam vents within a fumarolic area may indicate a gradation which can be related to reaction of individual components with wall rocks in shallow layers. Assuming progressive differentia-

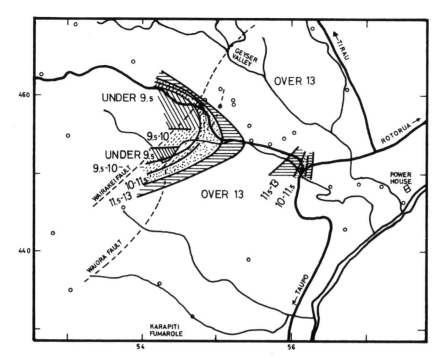

FIG. 5. Na/K ratios in water from drill holes in Wairakei, New Zealand (from Ellis and Wilson, 1960).

tion from the source to the surface, the chemical variations may indicate the area of major underground flashing. H_2S is probably the most sensitive indicator in this respect. Apart from the empirically proven association of the CO_2-H_2S-H_2-gas mixtures with higher underground temperature, the interpretation of gas analyses from natural fumaroles is difficult in detail and should be done with due regard to external conditions and the sampling technique since small variations in these factors may cause large spread in relative amounts of analysed components.

Volatile components such as ammonia and boric acid are present in some fumarole discharges, and become concentrated in the condensed water. Because of the relatively low volatility of boric acid and high reactivity of ammonia with near surface clay material, these components are not carried long distances in the steam phase. Where these components are present they may therefore show concentration gradation within the area which would indicate the zone of most intense subsurface boiling.

Underground gas-pressures, pH and scaling

Important factors in the operation of geothermal power plants are the amount of noncondensible gases associated with the steam, and the possibility of precipitation of solid substances in pipelines.

The amount of total gases in deep hot water varies within wide limits in different thermal areas. In Hveragerdi in Iceland it is 0.003 mole per cent, 0.02 mole per cent in Wairakei, New Zealand, and 0.07 mole per cent in the Mexicali field.

An exact figure on the gas/water ratio cannot be obtained except by sampling steam and water from a producing drill hole. Some indication can be obtained from measurements of gas/water ratios in fumaroles or steam-vents, but here the separation temperature is not known and gases may be lost by reaction with wall rock or absorption in shallow ground water.

Differences in total gas pressures between areas, which are otherwise near identical as to composition of dissolved solids, are probably related to difference in the amount of underground boiling and natural escape of steam at some stage in the history of the thermal field, rather than fundamental variations in the sources of chemical elements.

Ellis (1967) outlines the procedure for calculating underground pH from chemical analyses of water and steam from geothermal drill holes. The pH of the drill hole sample is different from the deep water due to concentration of the sample by boiling, loss of CO_2 and H_2S, and possibly by oxidation at the surface. All components, both water, steam and noncondensible gases, must be sampled at a known pressure and analysed. The results are subsequently computed back to a one-phase system at the known reservoir temperature. The drill hole must therefore draw its production from a liquid phase only, which can be verified by comparison of the directly measured drill hole temperature with the enthalpy of discharge.

The most significant pH-buffering systems are carbondioxidebicarbonate and silica-silicate, but other weak acid-base equilibria must be considered, such as hydrogen sulphide-bisulphide, boric acid-borate, ammonium-ammonia, etc.

Ellis (1967) compared two areas, Hveragerdi, Iceland, and Wairakei, New Zealand, and finds that the calculated reservoir pH is 6.6 at a temperature of 260 °C in Wairakei and 8.0 at a temperature of 216 °C in Hveragerdi. The directly measured pH in the drill hole discharges are 8.4 and 9.4 respectively.

If the changes occurring in the hydrogen ion concentration upon flashing of the water are known, it is possible to predict the possibilities of calcium-carbonate precipitation. This is an important factor in feasibility studies for power plants, since scaling calls for periodical clean-out operations. In severe cases calcite may precipitate in the formation outside the drill-hole and block feeding fissures. The drill hole is then damaged beyond repair. This can be controlled to some extent by pressure regulation at the valve, bringing the boiling into the hole.

Figure 6 (Ellis, 1969) is a phase diagram for 260 °C showing stability fields for hydrothermal alteration-minerals as a function of the ratios between 'K' and 'H' ions and 'Ca' and 'H' ions. Superimposed on the diagrams are lines showing the limiting conditions for Ca_2CO_3 precipitation at different molal concentrations of CO_2.

Precipitation of silica upon flashing of the thermal water seems to provide less trouble than calcite precipitation. The metastable persistence of silica in solutions, even at high supersaturations, was previously discussed in connection with the use of silica as a chemical thermometer. White, Brannock and Murata (1956) studied the behaviour of silica in hot spring waters and found that dissolved silica would polymerize quickly upon cooling, but stay in solution without precipitation for some time. Hot spring waters containing less than 200 ppm SiO_2 usually do not form sinter deposits.

Sampling of geothermal fluids

In order to obtain meaningful samples for chemical analysis of thermal water and gases, some factors, which may alter the composition of the sample on storage, have to be kept in mind. A good compilation of sampling methods and analytical procedures was published by Ellis and Ritchie (1961) and Mahon (1961, 1964). Only a few significant points will be mentioned here. Containers for water samples should preferably be made of polyethylene, since glass bottles could alter the silica and alkali content of the sample. If trace metals are to be determined in the water the sample container either absorbs the metal ions or contaminates the sample. An 'on the spot' precipitation of trace metals on a voluminous organic carrier is therefore necessary (Arnorsson, 1968). Bicarbonate and pH are preferentially determined in the field, but commonly special samples for these measurements are taken into pressure-sealed glass bottles for later analysis. Other volatile components, such as H_2S and ammonia are fixed chemically in the field.

The chemical analysis follows standard patterns, but special care must be taken with a few important components. Silica is usually determined colorimetrically with ammonium molybdate, but in supersaturated solutions silica tends to form polymers, which do not react with the molybdate. The sample aliquote is therefore digested with sodium hydroxide before analysis. Magnesium in low concentration is difficult to determine with the conventional EDTA complexometric titration, which usually results in high values. Because of the importance of magnesium as an indicator of high temperature, it is necessary to have an exact determination of this element. Atomic absorption spectometry is probably the most reliable method. In order to obtain meaningful values for the proportions of different weak acid-bases for calculation of underground pH, the alkalinity titration is performed after a method of Ellis and Ritchie

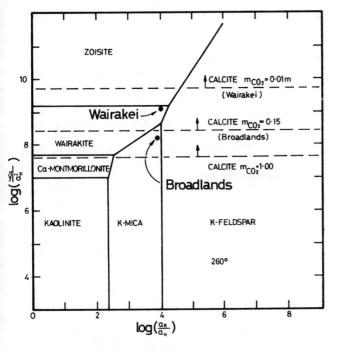

FIG. 6. Phase diagram for 260 °C showing estimated stability areas for various minerals at different ratios of 'a_{K^+}/a_{H^+}' and $a_{Ca^{++}}/a_{H^+}$.

Examples of natural thermal waters are shown and limiting values of '$a_{Ca^{++}}/a^2_{H^+}$' for calcite precipitation for different amounts of CO_2 in solution (1.00, 0.15 and 0.01 m CO_2). (From: Ellis and Wilson, 1960.)

(1961) which differentiates between the different carbonate species and the so-called 'effective bicarbonate', which is a measure of total alkalinity. Sampling of hot spring and fumarole gases is described by Ellis and Ritchie (1961), which also shows the general principle in sampling water

and gases from producing drill holes. Mahon (1966) describes a method for determination of enthalpy of discharge by measuring the amount of a gas component in the steam of two sampling points with different separation pressure on a blowing pipe.

Bibliography

ALONSO ESPINOSA, H. 1966. La zona geotérmica de Cerro Prieto, Baja California. *Bol. Soc. geol. mex.*, 29: 17-47.

ÁRNASON, B.; THEODORSSON, P.; BJÖRNSSON, S.; SAEMUNDSSON, K. 1969. Hengill, a high temperature thermal area in Iceland. *Bull. Volc.*, 33: 245-260.

ARNORSSON, S. 1968. *Dissertation*, London, Imperial College.

CRAIG, H. 1962. C^{12}, C^{13}, and C^{14} concentration in volcanic gases. *J. geophys. Res.*, 67: 1663.

——. 1963. The isotopic geochemistry of water and carbon in geothermal areas. In: E. Tongiorgi (ed.), *Nuclear geology in geothermal areas*, p. 17-53, Pisa, Consiglio Nazionale delle Ricerche, Laboratorio di Geologia Nucleare.

——; BOATO, G.; WHITE, D. E. 1956. The isotopic geochemistry of thermal waters. *Nuclear Processes in Geological Settings*, p. 29-44. (Natl. Res. Council Nuclear Sci. Ser. Rept. no. 19.)

DOE, B. R.; HEDGE, C. E.; WHITE, D. E. 1966. Preliminary investigations of the source of lead and strontium in deep geothermal brines underlying the Salton Sea geothermal area. *Econ. Geol.*, 61: 462-683.

ELLIS, A. J. 1959. The solubility of calcite in carbon dioxide solutions. *Amer. J. Sci.*, 257: 354-365.

——. 1961. Geothermal drillholes—Chemical investigations. *Proc. U.N. Conf. on New Sources of Energy. Rome, 1961* (E/CONF 35/G/42). New York, United Nations.

——. 1962. Interpretation of gas analyses from the Wairakei hydrothermal area. *N.Z. J. Sci.*, 5: 434-452.

——. 1963. The solubility of calcite in sodium chloride solutions at high temperatures. *Amer. J. Sci.*, 261: 259-267.

——. 1966. Volcanic hydrothermal areas and the interpretation of thermal water compositions. *Bull. Volc.*, 29: 575-584.

——. 1967. The chemistry of some explored geothermal systems. In: H. L. Barnes (ed.), *Geochemistry of hydrothermal ore deposits*, p. 465-514. New York, Holt, Rinehart and Winston.

——. 1968. Natural hydrothermal systems and experimental hot-water/rock interaction: Reactions with NaCl solutions and trace metal extraction. *Geochim. et cosmoch. Acta*, 32: 1356-1363.

——. 1969. Present-day hydrothermal systems and mineral deposition. *9th Commonw. Min. Metall. Congr. 1969.* (Mining and Petroleum Geology Section, Paper 7, p. 1-30.)

——; ANDERSON, D. W. 1961. The geochemistry of bromine and iodine in New Zealand thermal waters. *N.Z. J. Sci.*, 4: 415-430.

——; GOLDING, R. H. 1963. The solubility of carbon dioxide above 100 °C in water and in sodium chloride solutions. *Amer. J. Sci.*, 261: 47-60.

——; MAHON, W. A. J. 1964. Natural hydrothermal systems and hot water/rock interactions. *Geochim. et cosmoch. Acta*, 28: 1323-1357.

——; ——. 1967. Natural hydrothermal systems and experimental hot water/rock interaction (Part II). *Geochim. et cosmoch. Acta*, 31: 519-538.

——; RITCHIE, J. A. 1961. Methods of collection and analysis of geothermal fluids. (Report no. D.L. 2039.) 53 p. *Rep. Dep. Sci. industr. Res. N.Z.*, no. D.L. 2039. 53 p.

——; SEWELL, J. R. 1963. Boron in waters and rocks of New Zealand hydrothermal areas. *N.Z. J. Sci.*, 6: 589-606.

——; WILSON, S. H. 1960. The geochemistry of alkali metal ions in the Wairakei hydrothermal system. *N.Z. J. Geol. Geophys.*, 4: 593-617.

——; ——. 1961. Hot springs areas with acid-sulphate-chloride waters. *Nature, Lond.*, 191:696-697.

FACCA, G.; TONANI, F. 1964. Theory and technology of a geothermal field. *Bull. Volc.*, 27: 1-47.

GRINDLEY, G. W. 1961. Geology of New Zealand geothermal steam fields: *U. N. Conference on New Sources of Energy, Rome, 1961.* (E/Conf. 35/G/34.) New York, United Nations.

HEMLEY, J. J.; JONES, N. R. 1964. Aspects of the chemistry of hydrothermal alteration with emphasis on hydrogen metasomatism. *Econ. Geol.*, 59: 538-569.

MAHON, W. A. J. 1961. Sampling of geothermal drill hole discharges. *Proc. U. N. Conference on New Sources of Energy, Rome, 1961.* (E/CONF 35/G/46.) New York, United Nations.

——. 1961. Requirements for sampling Geothermal bore discharges for chemical analysis. *Rep. Dep. sci. industr. Res. N.Z.*, no. D.L. 2041.

——. 1962a. The carbon dioxide and hydrogen sulphide content of steam from drill holes at Wairakei, New Zealand. *N.Z. J. Sci.*, 5, 1: 85-98.

——. 1962b. A chemical survey of the steam and water discharged from drill holes and hot springs at Kawerau. *N.Z. J. Sci.*, 5: 417-432.

——. 1964. Fluorine in the natural thermal waters of New Zealand. *N.Z. J. Sci.*, 7: 3-28.

——. 1966. A method for determining the enthalpy of a steam/water mixture discharged from a geothermal drill hole. *N.Z. J. Sci.*, 9: 791-800.

——. 1967. Natural hydrothermal systems and the reaction of hot water with sedimentary rocks. *N.Z. J. Sci.*, 10: 206-221.

——; KLYEN, L. E. 1968. Chemistry of the Tokaanu-Waiki hydrothermal area. *N.Z. J. Sci.*, 11: 140-158.

MUFFLER, L. J. P.; WHITE, D. E. 1968. Origin of CO_2 in the Salton Sea geothermal system, Southeastern California, U.S.A. *23. Int. geol. Congr., Prague, Genesis of mineral and thermal waters*, vol. 17, p. 185-194.

——; ——. 1969. Active metamorphism of Upper Genozoic sediments in the Salton Sea geothermal field and the Salton Trough, Southeastern California. *Bull. geol. Soc. Amer.*, 80: 157-182.

ORVILLE, P. H. 1963. Alkali ion exchange between vapour and feldspar phases. *Amer. J. Sci.*, 261: 201-237.

SCHOEN, R.; EHRLICH, G. G. 1968. Bacterial origin of sulphuric acid in sulphurous hot springs. *23. Int. geol. Congr., Prague, Genesis of mineral and thermal waters*, vol. 17, p. 171-178.

SIGVALDASON, G. E. 1959. Mineralogische Untersuchungen über Gesteinszersetzung durch postvulkanische Aktivität in Island. *Beitr. Min. u. Petr.*, 6: 405-426.

——. 1963. Epidote and related minerals in two deep geothermal drillholes, Reykjavik and Hveragerdi, Iceland. *Prof. Pap. U.S. Geol. Surv.*, 450-E: 77-79.

——. 1966. Chemistry of Thermal Waters and Gases in Iceland. *Bull. Volc.*, 29: 589-604.

STEINER, A. 1963. The rocks penetrated by drillholes in the Waiotapu thermal area, and their hydrothermal alteration. *Bull. N.Z. Dep. sci. industr. Res.*, 155: 26-34.

UZUMASA, Y. 1965. *Chemical investigation of hot springs in Japan*. Tokyo, Tsukiji Shokan. 189 p.

WHITE, D. E. 1957a. Thermal waters of volcanic origin. *Bull. geol. Soc. Amer.*, 68: 1637-1658.

——. 1957b. Magmatic connate and metamorphic waters. *Bull. geol. Soc. Amer.*, 68: 1659-1682.

——; BRANNOCK, W. W.; MURATA, K. J. 1956. Silica in hot spring waters. *Geochim. et cosmoch. Acta.*, 10: 27-59.

——; HEM, J. D.; WARING, G. A. 1963. Chemical composition of subsurface waters. In: M. Fleischer (ed.) *Data of Geochemistry*, (6th ed.) Chapter F. (U.S. Geological Survey Prof. Paper 440-F.) 67 p.

——; ROBERSON, C. E. 1962. Sulphur Bank, California, a major hot spring quicksilver deposit. *Petrologic Studies; A Volume to Honor A. F. Buddington.* p. 397-428 *(Bull. geol. Soc. Amer.).*

WILSON, S. H. 1966. Origin of tritium in hydrothermal solution. *Nature, Lond.*, 211: 272-273.

The structure and behaviour of geothermal fields

Giancarlo Facca
Consulting Geologist
(Italy)

1. Introduction

Exploration data (geological, geochemical and geophysical) accumulated from a geothermal field can be used to build up a 'model', or picture, of the structure of the field and to explain its behaviour. In fact it is really only the interpretation of such data in the form of a model that enables important decisions to be taken as to the siting of drilling operations. All geothermal fields differ, of course, from one another. Nevertheless certain common features have enabled rational theories to be advanced concerning their structure and behaviour. Opinions still differ on many points of detail, but gradually a clearer understanding is emerging from the immense volume of exploratory and research work that has been, and is being, undertaken in many parts of the world.

In this article, geothermal fields are broadly classified and a basic model is postulated to account for their behaviour. Thereafter, a rough description is given of three particular fields in different parts of the world, to show how they appear generally to conform with the basic model.

2. Types of field

A primitive classification of geothermal field types, regarded as mining enterprises, distinguishes three broad classes, as follows:

2.1 HOT WATER FIELDS

Hot water fields, containing a water reservoir at temperatures ranging from 60 to 100 °C. Such fields can be useful for space heating, agricultural and various industrial purposes. The thermal gradients in fields of this type may range from 'normal' (about 33 °C/km), at which hot water of useful temperature would occur at depths from about 1,800 m to 3,000 m, to values of about twice the normal or more. Examples of the latter are the Hungarian Basin (50 to 70 °C/km gradient) and the Arzag Basin in Southern France (60 °C/km gradient), and many regions of the U.S.S.R. At these higher gradients the hot water is of course encountered at shallower depths.

The geology of hot water fields is much the same as for cold ground water systems. Systems suitable for commercial exploitation may be confined (artesian) or open (without a cap rock).

Hot water fields of this low temperature type are quite common. They may be worth investigating for commercial exploitation in areas where:

(i) a large water reservoir is believed to exist at temperatures of at least 60 °C at less than 2,000 m depth;
(ii) the heat flow is at least 2.2 μcal/cm²s (about 50% above world average);
(iii) the yield per well is large.

2.2 WET STEAM FIELDS

Wet steam fields, containing a pressurised water reservoir at temperatures exceeding 100 °C. This is the commonest type of economically exploitable geothermal field. Notable examples under exploitation are Wairakei (New Zealand), Cerro Prieto (Mexico), Reykjavik area (Iceland), the Salton Sea (U.S.A.) and Otake (Japan). The hottest known field is Cerro Prieto (380 °C).

When hot water is brought up a well to the surface, and its pressure is sufficiently reduced, some of the water will be flashed into steam, so that the resulting fluid is a mixture of water and steam under saturated conditions, with water usually predominating. The proportions of water and steam vary from field to field, and from well to well in a single field, according to the enthalpy of the fluid at depth and the pressure at the wellhead. A productive well in this type of field will continue to flow after the flowing process has been initiated. Such fields can be suitable for power generation, as well as for other purposes.

Giancarlo Facca

2.3 DRY STEAM FIELDS

Dry steam fields, or those that yield dry or superheated steam at the wellhead, at pressures above atmospheric. The degree of superheat may vary from 0 to 50 °C. Examples of this type of field under exploitation are Larderello and Monte Amiata (Italy), The Geysers (California) and Matsukawa (Japan). This type of field is also suitable for power generation and other purposes.

Geologically, wet steam and dry steam fields are generally similar, as emphasised by the fact that in some cases wells have produced wet steam for a period and dry steam later.

3. Basic model of a geothermal steam field

The basic features of a geothermal steam field, wet or dry, are shown in Figure 1. These features are:
 (i) a source of natural heat of great output;
 (ii) an adequate water supply;
 (iii) an 'aquifer', or permeable reservoir rock;
 (iv) a cap rock.

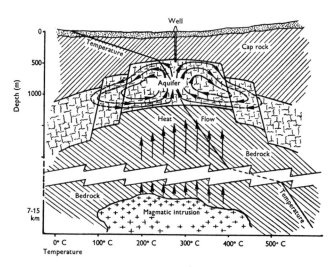

FIG. 1. Basic model of a steam field.

3.1 THE SOURCE OF HEAT

There is general agreement that the source of heat is a magmatic intrusion into the earth's crust, having a temperature of 600 to 900 °C, often at depths of the order of 7 to 15 km. This view is supported by various facts and reasons, notably that all the known 'commercial' fields are in regions where volcanic activity has occurred during recent Miocene-Quaternary times, or is still occurring. Some fields are actually situated on, or close to volcanoes (e.g. Japan and Central Mexico), whereas others (e.g. Larderello) are not directly linked with a centre of recent volcanic activity.

Larderello is nevertheless located within the northern part of the large Tyrrhenian volcanic province. Steam fields thus appear to be located in areas either of active, or of dormant (cryptovolcanic), activity.

In an active volcano, a magmatic intrusion reaches the surface through a large fault system. In compact, hard rock faulting may provide a channel for the upward flow of magma, while plastic rocks such as clay may flow by gravity into the fault space and seal it from above. The energy of a magmatic intrusion may be sufficient to penetrate the fault system in hard rocks but insufficient to prevail against the overburden of the plastic rocks. In such cases the magma may intrude to the boundary between the hard and plastic rocks. This cryptovolcanism may occur in areas devoid of recent volcanic activity, and is more likely to be found in geological areas of thick plastic formations, like a turbiditic series (flysch, greywacke). This applies to the two major dry steam fields, Larderello and the Geysers.

Magmatic intrusions without present eruption are common in acidic volcanoes and can also occur in basic volcanoes. Such intrusions provide the heat source for the Japanese and Central Mexican fields located on or around volcanoes.

Surface volcanic products like lavas, ignimbrites and tuffs cool too quickly to originate a 'commercial' geothermal field. Magmatic intrusions alone, that have occured within the last half million years or so, can satisfactorily account for the heat source: older intrusions would probably have cooled off by now. The problem is how to find evidence of 'recent' magmatic intrusions, some of which may be very deep seated.

In petroleum exploration it is recognised that the necessary geological environment for an accumulation of oil is a sedimentary basin where large quantities of organic matter have accumulated and evolved into hydrocarbons. The existence of such a basin, of whatever age, may offer the lithological and structural conditions required for the existence of a commercial oil field, and may therefore be regarded as 'prima facie' evidence of a possible field. Are there conditions that can equally be regarded as promising signs of the existence of a magmatic intrusion?

The occurrence of recent volcanism or the discovery of surface thermal manifestations are such signs, but are not absolutely essential. Abundance of mercury or arsenic are circumstantial evidence. Certain geological environments are also indicative of the possible existence of magmatic intrusions. This is a subject claiming the attention of vulcanologists and magmatologists. Two such environments are as follows:

(a) *Rift valleys and large grabens.*[1] The large structural depressions of the earth appear to be caused by the splitting apart of the crust at rates of several centimetres per year. The best known of these depressions in The Great East African Rift Valley, which passes from Tanzania through Danakil, The Red Sea, Jordan Valley, into Leba-

1. See fig. 6.

non and Syria. It is the site of Quaternary to active volcanism. Surface thermal activity is common in this valley but there could be good geothermal prospects in places where there is less obvious evidence of abnormal heat flow. Another area of this class passes through the Gulf of California and the Salton Sea, in North America. Professor Rex of the University of California is here carrying out a sophisticated programme of geothermal investigations that have revealed several promising geothermal areas many tens of kilometres from the nearest hot springs. The Basin and Range geological province in the Western U.S.A. is another such region: the basins are grabens.[1]

There are good reasons for believing that grabens[1] are the preferred site of magmatic intrusions, the chief of which is the fact that in a graben[1] the overburden is less than in the adjoining positive structures, so that magma can intrude to a level where its energy is balanced by the weight of the overlying stratigraphic series. Often there is a minimum of rock overburden in a graben,[1] so that the presence of cryptovolcanism is a definite possibility, even in the absence of hot springs.

(b) *Turbidites areas.* Larderello, Monte Amiata and The Geysers geothermal fields are all located in a turbidite geological environment. Similar geological features are extensively found elsewhere, and their closer investigation by geothermal geologists could perhaps be rewarding. Preliminary exploration has been carried out in one such region, the Flysch province of North-East Algeria. Here only one very small volcanic centre is known, and surface manifestations of anomalous heat flow are very limited. Nevertheless the general structure and stratigraphic column are similar to those of Larderello, and several granitic intrusions of the Miocene age are known. Geochemical data and a few shallow gradient bores indicate a fairly high temperature gradient. Hitherto, these investigations have proceeded no further.

3.2 WATER SUPPLY

Early hypotheses about geothermal fluids suggested that they were of 'magmatic', or 'juvenile' origin, that is, water vapour and gases released from solution in the magma when the pressure is reduced. Whilst this may still be partially true it is now believed that at least 90% of the water in a geothermal reservoir is 'meteoric', originating from rain water. Geochemical evidence based on isotopic measurements, particularly that offered by Craig, Boata and White in 1956 (Craig *et al.*, 1956) support this opinion. Thus, referring to Figure 1, it would appear that most of the water in the aquifer is of meteoric origin and that it is heated conductively through a largely impermeable base rock, even though some relatively small quantity of magmatic steam may penetrate this base rock through faults and fissures. This theory was first advanced by Goguel in 1953 (Goguel, 1953).

As hot fluid is withdrawn from bores or from surface vents, the hydrological balance of the system is restored, or partially restored, by the inflow of new water. There are often clearly visible 'recharge areas', where the permeable reservoir terrain outcrops, permitting the ingress of rainwater. At Larderello there are outcrops of the Mesozoic carbonatic and evaporitic series. Here, it is possible to calculate from observed rainfall and run-off the quantity of meteoric water entering the reservoir each year. In other geothermal fields matters are not always so simple. Many hydrothermal systems are dynamic, with water entering at some high level and leaving at some low level. Our knowledge of water movements in deep aquifers is very limited especially where these lie below sea level, as is so often the case. Further investigations in this direction will have to be made.

3.3 THE AQUIFER

A good productive geothermal well should produce at least 20 t/h of steam: many wells produce a great deal more. A 'wet' well may produce some hundreds of tons per hour of mixed fluid. The maintenance of such high flow rates implies a high degree of permeability in the aquifer, with porosity playing only a secondary part. Any permeable rock can serve as a good geothermal reservoir. At the Geysers it is greywacke with fissure permeability; at Larderello a carbonate rock with karstic permeability; at Wairakei, fissured ignimbrite overlain with rhyolite and pumice breccia; at Otake, a permeable volcanic tuff; at Cerro Prieto, deltaic sands.

3.4 CAP ROCK

A cap rock is a layer of rock of low permeability overlying the aquifer. All steam producing fields have a cap rock. Some have been formed as original impervious rocks, such as the flysch formation at Larderello, the lacustrine Huka formation at Wairakei, the deltaic clay in the Imperial Valley and Cerro Prieto fields. Elsewhere, the cap rock may have become impervious as a direct result of thermal activity. For example, at The Geysers and at Otake, the shallow rocks are hard fractured formations. It is probable that before the beginning of thermal activity these rocks had a fissure permeability, but that this activity itself has caused the sealing of the permeable passages. This could have occurred by two geochemical processes:

(a) the deposition of minerals from solution, mainly silica;
(b) the hydrothermal alteration of rock, causing kaolinisation.

This theory was advanced by Facca and Tonani in 1964 (Facca, 1969; Facca and Tonani, 1964*a*, 1967). Deposition of silica can easily be observed in The Geysers field: fractures of one-inch width, completely filled and sealed by silica and calcite are common features. Kaolinisation, associated with other more complicated hydrothermal rock

---

1. See fig. 6.

alteration, is also widespread and prominent. Hydrothermal alteration can be recognised by the bleaching of greywacke, and in places by the lack of vegetation. Hydrothermal alteration is a very complicated, and not fully understood, geochemical process that changes from place to place.

Sometimes a hot aquifer may intercalate in places with the cap rock. Where such zones have been exploited they have been productive only for a limited time and it has been necessary to deepen the bores into the main convective reservoir.

4. Basic model of a low temperature hot water field

A low temperature hot water field (para. 2.1 above) may sometimes occur in an environment similar to that shown in Figure 1. It can also occur in fields devoid of cap rock, in which case the model is conceived somewhat in the manner of Figure 2, where the thermal gradient and depth of the pervious aquifer are sufficient to maintain a convective circulation. The temperature in the upper part of the reservoir will not exceed the boiling point at atmospheric pressure, partly because water brought up convectively from depth will lose pressure (and temperature) as it rises, and partly because there may be mixing with cool ground waters.

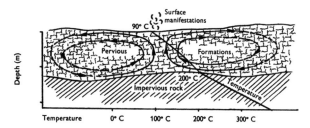

Fig. 2. Basic model of a low temperature hot water field.

5. Convection and steam formation in a steam field

If the permeability and depth of the aquifer are sufficient, and the supply of deep heat also sufficient, a convection system will be set up and maintained (see Fig. 1). Thus hotter water will be conveyed from the lower to the upper part of the aquifer and temperatures will tend to stabilise at or near the boiling point corresponding to the hydrostatic pressure, in the rising stream of the convective system which forms the active part of the field. Nevertheless, temperature inversions are not uncommon owing to secondary disturbances in the convective system. A reservoir may be wholly filled with water—often at or near boiling point corresponding to depth—or it may contain an upper layer of steam and gases.

When a bore is sunk into the water zone and fluid passes up it flashing will occur owing to the reduction of pressure. This flashing may occur:
 (i) in the well itself;
 (ii) at the walls of the well where water enters it;
 (iii) within the reservoir at some distance from the well.

The location of the point of flashing depends on the temperature and the permeability of the reservoir. In a 'wet' geothermal field, flashing occurs at (i) or possibly at (ii). In a dry field it occurs within the reservoir. It takes place, moreover explosively, so that rock pieces (as big as a fist) are erupted. At the Larderello and Geysers fields, where no bore liners are provided, much rock and dust are erupted when a bore is first 'blown'.

If the water recharge rate is insufficient to balance the draw-off of steam, an 'evaporation space' will form round the base of a well. This space will grow in size and may eventually join the evaporation spaces of other producing wells, so that a comprehensive steam zone may form above the water reservoir. Evidence from certain geothermal fields shows that the steam/water interface, in such cases, tends to fall if production is high. This process is likely to be assymptotic, owing to the higher rate of recharge as the level falls. A point of equilibrium is likely to be reached eventually.

6. Superheat

The occurrence of superheated steam at some fields (e.g. Larderello and The Geysers) has caused some speculation. Some theorists maintain that this is evidence of the magmatic origin of the steam. In 1953 Goguel suggested it was due to thermal conduction at the base rock/aquifer interface (Goguel, 1953). In 1961 Facca and Tonani suggested that it could be due to the throttling action on saturated steam passing through the reservoir (Facca and Tonani, 1964a). (Unknown to them, a similar theory had been advanced in 1924 by L. H. Adams (1924).)

7. Mineral deposition

As water flashes into steam at the water/steam interface of a dry field, some cooling takes place. This can result in the deposition of minerals, notably silica (as described in Section 3.4 above), and slowly lead to reduced permeability and lower steam output. Even in a wet field, the lowering of the water table, as a result of draw-off, can cause a fall of pressure and temperature, with consequent mineral deposition. This may account for the commonly observed decline in well outputs even where no deposition is found in the actual bores.

8. Surface manifestations

An abnormally high heat flow often, though not always, gives rise to surface manifestations, such as:

- warm or hot springs, steaming ground, steam vents and geysers;
- hydrothermal rock alterations;
- silica and travertine sinter;
- abnormally warm ground water temperature;
- fumaroles and solfatares;
- mercury ores of recent origin;
- hot soils, sometimes revealed by anomalous snow melting.

It should be remembered that thermal anomalies can exist without associated surface manifestations (e.g. Monte Amiata) and also that such manifestations have sometimes failed to lead to a commercially exploitable field.

Hot springs can originate from:
- deep circulation and resurgence through a permeable fault in an area of normal, or near normal, heat flow;
- local exothermic geochemical phenomena, such as anhydrite to gypsum or sulphide oxidation;
- heating by lavas not yet cooled, or by the molten magma of volcano;
- heating by magmatic intrusion.

Springs of the first type may be indicative of a hot water field of the type described in Section 2.1 above: those of the second and third types are more or less worthless economically: those of the fourth type are indicative of good economic prospects.

Certain types of hot spring or steam vent have been described by Tonani as 'leakage manifestations' (Tonani, 1970). He suggested that water boiling at depth in a deep reservoir could pass up to the surface through a fault (often an active one) much as it would pass up a well. This water would flash into steam and may condense by cooling near the surface. The condensed water would differ chemically from normal surface water and its unusual character could also be detected after admixture with surface water. Sometimes there can be a leakage of gas, from the upper zone of a reservoir, through a fault to the surface, where it heats and mixes with ground water, thus giving rise to a hot spring rich in gases. An analysis of these gases can enable the temperature of origin to be deduced.

Hot springs having a high discharge rate of water may imply an aquifer at the same, or slightly higher, temperature.

Many hot springs deposit impressive siliceous and travertine sinter, which may be large and thick. Siliceous sinter deposits are indicative of a high-subsurface temperature at the time of deposition: travertine deposits may imply low temperatures.

9. The Geysers geothermal field, California

9.1 GENERAL

The Geysers steam field is located in the Coast Range geological province of California, about 140 km north of San Francisco. It is one of the few dry steam fields now being exploited. With an impressive record of successful drilling (70 producing wells in 1970, in a total of 75 drilled) the field is still under development. Several additional bores are being sunk every year and the boundaries of the field have not yet been reached. The field would appear to be the largest in the world. Steam equivalent to about 400 MW of power has already been proven; while the potential of the drilled area, which covers at least 20 sq. miles, has been conservatively estimated at 1,300 MW. There are reasons for believing that the exploitation of the whole thermal area could produce at least 3 or 4 million kilowatts.

9.2 HOT SPRINGS

Hot springs and steam vents occur in the local areas known as 'The Geysers', 'Sulphur Bank' and 'Happy Jack'. It was in these areas that the first bores were sunk. Other local areas such as 'Squaw Creek', 'Tyler Creek' and 'Castle Rock' have either no hot springs at all or few warm springs of low discharge. Nevertheless, steam has been produced in good quantity in some of these areas. This shows that hot springs alone cannot be regarded as necessary indications of a useful steam field. The extent of the entire thermal area is thus judged to be far larger than was at first thought on surface evidence. In any case, hot springs are transient phenomena that can disappear. Thus Anderson hot springs disappeared at the time of a landslide in 1969, and in another thermal area Byron hot springs dried up as a result of a lowering in the water table.

9.3 GEOLOGY

Widespread young volcanic activity is an outstanding geological feature of the region. At the centre of the field the Cobb Mountain is a rhyolitic dome. At the present eastern end of the field, in the Middletown area, there are very young volcanic vents, lava flows and tuffs; also the St. Helena Mountain volcanics of the Sonoma series. Northwards there are Mounts Hanna and Konocti which are very recent volcanic domes.

The reservoir is provided by the Franciscan greywackes, and sometimes by related rocks (serpentinite). Although the primary permeability of greywacke is almost nil, intensive fracturing provides high secondary permeability. When drilling, circulation losses are common, especially below 3,000 ft. The reservoir appears to be very thick, certainly over several thousand feet and possibly more than 10,000 ft. This could provide ideal conditions for deep seated fluid convection.

At shallow depth there is a layer of impervious rock, as proven by air drilling. Although some water bearing layers of limited thickness have been penetrated by some wells, the series as a whole is impervious and can thus provide the 'cap rock' as described in Section 3.4 above. Much of this rock appears to be an example of the self-sealing process described in that section: elsewhere it is formed of primary siltstone.

9.4 FIELD MODEL

In the early days of exploration in the Geysers field, the theory was commonly held that the origin of the steam was magmatic. It was believed that juvenile steam travelled upwards directly from the magma through fault zones. According to this theory the production of superheated steam should occur only along the fault zone. However, intensive drilling has shown that many good steam producing wells have been sunk in areas where no faults have been detected, that at least one of the few 'failure' wells has been drilled along a fault, and that the steam temperature and chemistry is similar for all wells. These facts, together with the geochemical evidence referred to in Section 3.2 above, have led to a modified theory that either none or only small proportion of the total field steam (less than 10% according to isotope tests) originates in the magma.

A great deal of exploratory work—geological, geochemical and geophysical—has been undertaken by the author and others during the last 7 or 8 years. A synthesis of all the data collected has enabled a tentative model, or picture, to be built up of the Geysers field. This model, shown in simplified form in Figure 3, has enabled predictions to be made as to the existence of promising drilling sites outside the confines of what was formerly believed to be the field, with excellent practical results. It has also led to a new evaluation of the power capacity of the field, far exceeding earlier estimates.

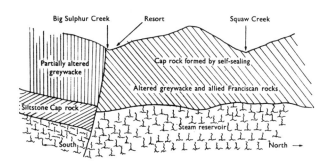

FIG. 3. Simplified model of the Geysers Field, California.

An important vertical fault will be noted in Figure 3. This runs from East to West beneath the Big Sulphur Creek for some distance and then diverges to the North-West. It was detected by lithological differences between its northern and southern walls. Northwards, the greywacke predominates from top to bottom, while to the south there is a well defined siltstone zone, some 2,000 ft thick, at a depth of about 3,500 ft. This siltstone provides the cap rock in the southern part of the field, while self-sealed greywacke and allied Franciscan rocks from the cap in the northern part of the field. The steam reservoir extends to a great depth, well below the deepest bore, and is believed to be heated mainly by conduction through bedrock from a magma intrusion beneath. Possibly a moderate amount of juvenile steam also ascends through the bedrock into the reservoir. The upper part of the reservoir, after many years of production, is now filled with steam and gases, while deeper down it is believed to contain boiling water. The extent of the field is probably as great in area as that of the magmatic chamber at depth: its size is not yet known.

This model may become modified in detail as more exploration data are accumulated, but it seems to be basically rational and explains the behaviour of the field.

10. The Otake geothermal field, Japan

10.1 GENERAL

Japan is a country of great geothermal possibilities, the total potential of the country having been tentatively estimated at several millions of kilowatts. The Otake field, here described, is located in Kyushu Island, a few kilometres north-west of Mount Kujyu-Zan, the highest peak in the Kujyu volcano group: it lies about midway between the Aso caldera[1] and Beppu Spa. A 13 MW power plant was commissioned in 1967 and has since been operating at high plant factor. It is planned to develop the area to the extent of about 180 MW.

10.2 SURFACE ACTIVITY

The Otake field lies in a faulted caldera[1] basin surrounded by volcanoes. Numerous hot springs and fumaroles located along the major faults in the basin indicate abnormally high heat flow. Alteration zones are also prominent.

10.3 STRATIGRAPHY

A great deal of sophisticated field exploration has been undertaken in the Otake field. Geological and geochemical surveys have been supplemented with geophysical work, including resistivity, magnetic, gravimetric, seismic, thermal gradient and heat flow measurements. As a result of this work, geological strata in the field have been identified and grouped in three complexes, so:

Complex 1 (uppermost), the Kuju volcanic Upper Pleistocene complex. This is mainly composed of horneblende andesite containing augite or hyperstene, intercalated with pumice flow of the same rock type in the middle section. The complex is fractured and hydrothermally altered, and has fissure permeability.

Complex 2 (middle), the Hobi volcanic Lower Pleistocene complex. This layer merges almost imperceptibly with Complex 1. It consists of the alternation of two-pyroxene lavas and their tuff breccias. The upper part of this complex is permeable, but the lower part impermeable. The variation in permeability is due to differences in the degree of hydrothermal alteration.

Complex 3 (lowest), of Miocene sediments. It consists of volcanic pyroclastics and its depth is more than 1,000 m. The complex is permeable and is the producing horizon of the field.

1. See fig. 6.

10.4 TENTATIVE GEOLOGICAL HISTORY OF THE FIELD

Volcanic activity began in the Miocene age and continued till the Holocene. In the Miocene age the area became covered with the Complex 3 rocks. Later, a deep magma body intruded below. Fracturing and fissuring occurred, imparting secondary permeability to the Complex 3 rocks. Meanwhile, the surrounding volcanoes became active again and covered the area with the lavas and tuff breccias of Complex 2, thus providing a cap rock over the Complex 3 convective system, the temperature of which in consequence rose. A new cycle of volcanic activity and faulting caused the deposition of the Complex 1 rocks and at the same time shattered the Complex 2 rocks. Hot fluids, previously con-fined in the convection system, migrated to the surface through active faults. In doing so they caused alteration in the upper strata and also mineral deposition, thus forming a cap rock of Complex 2.

10.5 FIELD MODEL

Figure 4 shows a cross section of the Otake field. The most productive bores, H-1 and H-2, are drilled in the highest fault block in the field. Thus the field broadly follows the conventional model, with a magmatic intrusion at the base, an aquifer (Complex 3) and a cap rock (Complex 2). The thermal fluid is mainly meteoric, though it is possible that a small proportion of magmatic fluid may be present.

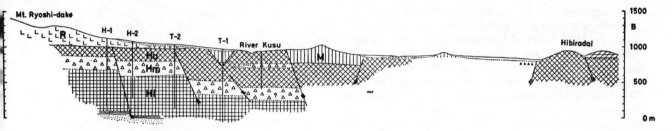

FIG. 4. Geological cross section of the Otake geothermal area.
Key: Hu = Upper formation
Hm = Middle formation
Hl = Lower formation
R = Ryoshi-dake lava
M = Misokobushi-yama lava
H-1 ⎫
H-2 ⎬ Production bores
T-1 ⎫
T-2 ⎬ Test bores
(From: Yamasaki, Y. Matsumoto and M. Hayashi, 1970.)

11. The Larderello geothermal field, Italy

11.1 GENERAL

This field, occurring in Tuscany about 50 km south of Pisa, is at present being exploited to a greater extent than any other geothermal field in the world. About 390 MW of power plant is generating about 3 million kWh per annum. The field yields superheated steam, sometimes reaching 260 °C. Typical conditions are 5 ata/225 °C at the well-head; maximum enthalpy 705 kcal/kg. Well yields generally decline with time, and the total power output has for some time been maintained by drilling new wells to make good the loss of output from the older wells. Hot springs and steam vents abound in the area.

11.2 GEOLOGY

The field lies in the southern part of the Era graben.[1] The stratigraphic series is as follows:

Group 1. Clay, sands, conglomerates. Pliocene, Upper Miocene.
Group 2. Flysch formation: shales, marls, limestones, ophiolites. Eocene, Upper Jurassic.
Group 3. Sandstones and shales. Oligocene, Cretaceous.
Group 4. Radiolarites, marls, cherty limestones and magnesium limestones with anhydrite layers. Upper Jurassic-Upper Triassic.
Group 5. Schistose-quartz formation (basement). Upper Triassic-Upper Carboniferous.

Groups 2 and 3 constitute the cap rock: groups 4 and 5 the reservoir. Some permeable beds of groups 2 and 3 produced small quantities of steam in early times. The limestones of the main reservoir have very high karstic and fissure permeability. The reservoir thickness varies, sometimes reaching several hundreds of metres.

1. See fig. 6.

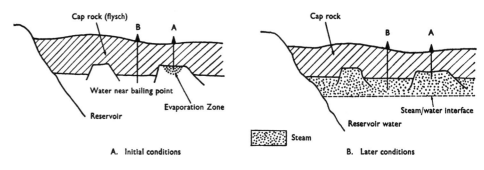

FIG. 5. Probable evolution of Larderello field.

11.3 TENTATIVE EXPLANATION OF FIELD BEHAVIOUR

It is suggested that before steam production began, the reservoir was filled with liquid water near boiling point. When fluid was drawn off from a well penetrating a reservoir dome (bore 'A', Fig. 5-A) an 'evaporation space' would be formed in the manner described in Section ·5 above. A bore penetrating into a deeper part of the reservoir (bore 'B' in Fig. 5-A) would have emitted flashing hot water. In the course of time the evaporation spaces of the shallower wells would have become enlarged until they merged into one another, eventually producing a condition as shown in Figure 5-B, in which all the upper part of the reservoir is filled with steam (and gas) and the lower part filled with water. All wells would then produce steam only. The steam/water interface is believed to have fallen continuously owing to the insufficiency of recharge water to make good the steam draw-off. At the same time, the flashing of water at the steam/water interface would give rise to mineral deposition, with consequent reduction in permeability and therefore of steam production.

This theory accords with the actual behaviour of the Larderello field, and although further confirmation from appropriate petrological and geochemical investigations is still awaited, strong evidence in support of the theory has been obtained by Professor Marinelli from cores and cuttings of the Larderello wells (Marinelli, 1970).

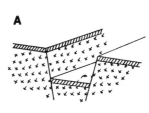

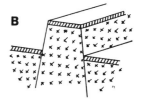

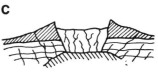

FIG. 6. Some important earth structure forms. A. A '*graben*' is a longitudinal down-faulted block, formed by complex stresses in the earth's crust; B. A '*horst*' is a longitudinal up-faulted block, formed by complex stresses in the earth's crust. It is the converse of a '*graben*'; C. A '*caldera*' is a depression of more or less circular form, formed by volcanic collapse, as a result of withdrawal of magma from beneath; usually by surface eruption elsewhere, but sometimes by migration of the magma within the earth.

Bibliography

Note. The reference 'Rome' denotes the United Nations Conference on New Sources of Energy, held in Rome in 1961 and published by United Nations, New York, 1964, as Proceedings of that Conference.

The reference 'Pisa' denotes the United Nations Symposium on the development and utilization of geothermal resources, held in Pisa in 1970, the proceedings of which have been published by the Istituto Internazionale per le Ricerche Geotermiche, Lungarno Pacinotti 55, Pisa, Italy.

ADAMS, L. H. 1924. A physical source of heat in springs. *J. Geol.*, vol. XXXII, no. 3, p. 191-194.
BODVARSSON, G.; PALMASON, G. 1961. Exploration of subsurface temperature in Iceland. Rome G/24.

CRAIG, H.; BOATO, G.; WHITE, D. E. 1956. The isotopic chemistry of thermal waters. Natl. Research Council, Nuclear Science. Report no. 19: Nuclear processes in geological settings. Publ. 400.

FACCA, G. 1969. Geophysical investigations in the self-sealing geothermal fields. *Bull. Volc.*, vol. XXXIII, p. 119-122.

——; TONANI, F. 1961. Natural steam geology and geochemistry. Rome, G/67.

——; ——. 1964. Theory and technology of a geothermal field. *Bull. Volc.*, vol. XXVII.

——; ——. 1967. The self-sealing geothermal field. *Bull. Volc.*, vol. XXX, p. 271-273.

GOGUEL, J. 1953. Le régime thermique de l'eau souterraine. *Ann. Min., Paris*, vol. X, p. 3-32.

——. 1970. Le rôle de la convection dans la formation des gisements géothermiques. Pisa.

GRINDLAY, G. W. 1964. Geology of New Zealand geothermal steam fields. Rome G/34.

HAYASHIDA, T.; EZIMA, Y. 1970. Development of Otake geothermal field. Pisa.

McNITT, J. 1961. Geology of the Geysers thermal area, California. Rome G/3.

MARINELLI, G. 1970. Floor discussion. Pisa.

NOGUCHI, T.; NISHIKAWA, K.; ITO, I.; USHIJIMA, K. 1970. Some theoretical considerations on hydrothermal systems due to cracks. Pisa.

TONANI, F. 1967. Some geochemical criteria in geothermal exploration. IV Symp. on geothermal problems. (I.H.F.C.) *XIV. I.U.G.G. Gen. Ass. Zürich.*

——. 1970. Geochemical methods of exploration for geothermal energy. Pisa.

YAMASAKI, T.; MATSUMOTO, Y.; HAYASHI, M. 1970. Geology and hydrothermal alteration of Otake geothermal area— Kyuji volcano group, Kyushu, Japan. Pisa.

III

The winning of geothermal fluids

Drilling for geothermal steam and hot water

Keiji Matsuo

President of Teiseki Drilling Co.
Geothermal Energy Association, Tokyo (Japan)

1. Introduction

Geothermal steam is sometimes found at depths as shallow as 50 m to 200 m, but the output from such sources is often impermanent, and the steam used for power or industrial purposes is usually found by drilling to depths of 500 m to 2,000 m. Exploration for geothermal steam sometimes necessitates drilling as deep as 2,500 m to 3,000 m, but this is rather unusual.

Geothermal wells are generally drilled with standard rotary rigs (Fig. 1) as used for crude oil and natural gas, of capacities necessary to reach the required depth. In some respects, however, drilling for steam differs considerably from that for oil and gas, as will be seen from what follows.

2. Drilling rig

As already mentioned, the rigs used are normal oil well rigs, and many varieties are in use throughout the world according to the depth required. The sizes and ratings for typical rigs as manufactured by the National Supply Company are given in Table 1.

3. Method of drilling

The method used is the usual rotary drilling, either with mud or with air. The former is the most commonly used but the latter, being faster and cheaper, has attracted increasing attention in recent years.

TABLE 1. Sizes and ratings of drilling rigs

Drilling depth	feet	1,500 to 3,000	2,000 to 4,000	3,000 to 5,500	4,500 to 7,500	6,000 to 9,000
	metres	500 to 1,000	700 to 1,300	1,000 to 1,800	1,500 to 2,500	2,000 to 3,000
Draw works	type	T-12	T-20	T-32	T-45	T-55
	horsepower	200	300	400	550	700
Pump	type	C.250. $7\frac{1}{4}'' \times 15''$ 1 set	C.250. $7\frac{1}{4}'' \times 15''$ 2 sets	K.500. $7\frac{1}{2}'' \times 15''$ 1 set C.250. $7\frac{1}{4}'' \times 15''$ 1 set	K.500. $7\frac{1}{2}'' \times 15''$ 2 sets	G.700. $8'' \times 14''$ 2 sets
	horsepower	370	370 × 2	513 370	513 × 2	700 × 2
Derrick	type	standard	standard	standard	standard	standard
	height	27″ or 34″	34″ or 38″	38″	38″	43″
Rotary table		$17\frac{1}{2}''$	$17\frac{1}{2}''$	$20\frac{1}{2}''$	$20\frac{1}{2}''$	$20\frac{1}{2}''$

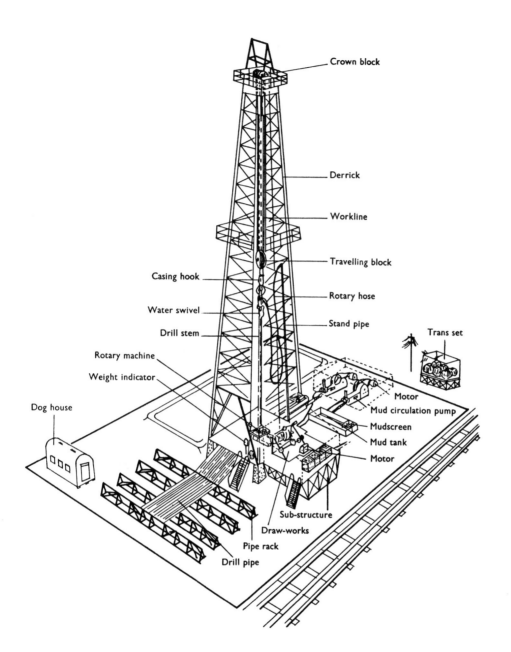

FIG. 1. Rotary drilling rig.

4. Drilling method with mud circulation
(see Fig. 1)

4.1 DRILLING PROCESS

Ground formations in geothermal areas consist mostly of volcanic rocks, characterised by a high hardness index, a high temperature gradient and strong faulting and fissuring. Losses of circulating fluid are therefore frequent and progress may be much slower than when drilling for oil or natural gas.

The time required for drilling is greatly influenced by the topography at the site, the ease of access, the hardness of the formations, the efficiency of the rig, the amount of casing required, the lengths to be cored and the number of metres to be drilled in the productive zone where mud losses are high. Under favourable conditions the following times are reasonable estimates:

Depth	Actual drilling	Finishing and testing
500 m	15 to 30 days	10 days
1,000 m	25 to 45 days	10 days
1,500 m	35 to 55 days	10 days
2,000 m	50 to 70 days	10 days

4.2 CASING PROGRAMME FOR STEAM PRODUCTION WELLS

An important item in the planning of a steam production well is a proper casing programme. A large volume of steam production should be expected as a matter of course from a large diameter well. But if a well is too large for the capacity of the steam formation it will not always produce abundant steam and is more likely to be incapable of maintaining a sustained flow. Furthermore, if casing has not been put down to the correct depths, and hot water occurs in higher formations, it may prove impossible to obtain a continuous flow of steam from the lower formation because of the incursion of hot water from above.

Further problems arise from the pH values of the hot water produced with the steam. If the pH value of the hot water is low, the casings will be heavily corroded and the life of the well will be shortened accordingly and may even become uneconomic.

In some cases slotted liners can be used, but it is best to avoid them unless there is risk of the hole walls collapsing during the long period of steam production. Slotted liners should be inserted if particles sloughed off the sides of the hole are recognised amongst the drill cuttings when passing through the production zone. The well can be completed without inserting a slotted liner, and sloughed particles can be watched for when letting the steam flow: the slotted liner can then be inserted if necessary. If, however, the hole walls collapse, it will require many days to re-drill, and it is safer to insert a slotted liner at the outset.

The casing programme will also differ according to the volume of hot water produced with the steam. As a general rule it will conform with the following classification:

(a) Steam volume 10 to 25 t/h:

17-in open hole	13 3/8″ surface casing	fully cemented
12¼-in open hole	9 5/8″ intermed: casing	fully cemented
8 5/8-in open hole	7″ production casing	fully cemented
6¼-in open hole	4½″ slotted liner	

(b) Steam volume 25 to 50 t/h:

18-in open hole	16″ surface casing	fully cemented
14¾-in open hole	11¾″ intermed: casing	fully cemented
10 5/8-in open hole	8 5/8″ production casing	fully cemented
7 5/8-in open hole	6 5/8″ slotted liner (O.D. of coupling skimmed by 1/16″)	

(c) Steam volume 50 to 80 t/h:

22-in open hole	18″ surface casing	fully cemented
17-in open hole	13 3/8″ intermed: casing	fully cemented
12¼-in open hole	9 5/8″ production casing	fully cemented
8 5/8-in open hole	7″ slotted liner	

4.3 CASING

Casing to be used for the production of large volumes of steam from subterranean formations must be able to withstand vibration, attrition through friction, wear and corrosion so as to remain in service for as long as possible. To minimise the attrition of the slotted liner it must be as large as possible by comparison with the hole diameter. For instance, in an oil well a 5½-in casing is inserted into an open hole of 7 5/8-in diameter: whilst in a geothermal well a 6 5/8-in casing (with the outside diameter of the coupling skimmed by 1/16-in) will be used in a hole of the same diameter. This is to ensure a minimum clearance between the outside of the casing and the walls of the hole. The sectional area of the 6 5/8-in casing is 40% greater than that of the 5½-in casing. This not only results in a greater volume of steam from a fertile reservoir, but also reduces the pressure losses within the casing, the vibration, shocks producing attrition, and wear, thus extending the life of the well.

As regards the wear of the casing, steam produced at a rate of 70 t/h passing through an 8 5/8-in CP nearly reaches sonic speed, which severely damages the casing.

The steam can come out superheated, dry saturated or wet. Steam with a wetness of 20 to 30% or more is the most damaging. A film of water, some 0.03″ thick, appears to cover the inside walls of the casing. Severe damage is caused when this film is removed by high speed steam.

Upon completion of a well the casing is sometimes worn by the fine sand carried up with the steam. This can be controlled to some extent by attaching a flow regulator (choke) to the wellhead in order to limit the steam speed.

As regards the corrosion of casings, it is nearly always caused by H_2S and other acids. Formations yielding water of low pH value, as determined by sampling (if possible) during the drilling of the well, should be cemented off. Alternatively they should be sealed off by casing which is cemented off to prevent the access of low pH water into the well.

As regards the casing itself, there is usually a difference of diameter at the inside of the coupling or at the bell collar of the inner pipe. This produces turbulence in the high speed steam flow and wears the upper corner (Fig. 2-A) and the inside surface (Fig. 2-B) of the casing. Such damage to the casing can be avoided by using a flush butt joint

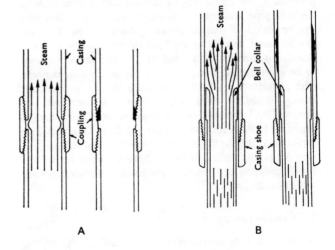

Fig. 2. Casing erosion at joints.

	Unit	Temp. °C × H time[1] 150 °C × 16 H	Temp. °C × H time[1] 200 °C × 16 H
AV (apparent viscosity)	c.p.	12	15
PV (plastic viscosity)	lb/100 ft²	10	13
YV (yield value)	ditto	4	4
Gel (gel strength)	ditto	1 to 10	1 to 16
WL (water loss)	c.c.	9.0	10.2
pH value		10.5	10.5

1. H time = heating time (hour).

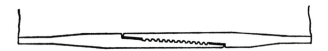

FIG. 3. Internal flush butt joint.

(Fig. 3), and steam turbulence at the bell collar can be avoided by tapering the bell collar section so that its inside diameter no longer changes suddenly from that of the production casing.

J55 steel to A.P.I. standards is generally used in most countries for the manufacture of casings, but for cases of serious corrosion it is essential to use an acid-proof casing capable of resisting hot, highly acid geothermal waters.

4.4 WELLHEAD EQUIPMENT

The first problem to be settled is whether to provide a cellar below the drilling rig. This is convenient to house the sub-base below the derrick and the blow-out preventers, as well as to provide better access to the valves during production. For security, however, it is better not to provide a cellar because of leakages which may occur in the wellhead equipment and the difficulty of subsequent repairs.

Wellhead equipment comprises double gate blow-out preventers, the upper stage equipped with a ram to hold the drill-pipe and the lower stage with a blind ram that can be sealed off without affecting the drill-pipe. The ram should be of natural or artificial rubber capable of withstanding temperatures around 150 °C and so constructed to be water-cooled if the hole is sealed off for long periods. The preventers are held by a flange on the intermediate casing. When the hole is completed, the preventers are removed and replaced by the main valve. Casing flanges are made to ASA standards in the 150, 300, 400 and 600 psi series according to the closed-in pressure required.

All piping at the wellhead above the main valve must be streamlined and the wall thicknesses increased to resist corrosion and wear.

4.5 DRILLING MUD

At comparatively low temperatures, around 150 °C, drilling of geothermal wells is usually carried out with bentonite or other clay-based muds. Above this temperature the mud tends to 'gel' and the filtrate to increase. CL-CLS based mud is therefore generally used with satisfactory results.

CL-CLS denotes chrome-lignite/chrome lignosulphonate. This type of mud is very effective in formations which easily cave in, and when cement contamination is pronounced. It is stable at high temperatures and can withstand up to 250 °C. Its composition and characteristics are as follows:

Composition:

Bentonite	8 to 9%
Blended ligno-sulphonite compounds	4 to 5%
Sodium hydroxide	0.1 to 0.15% pH 9.3 to 9.5
De-foamer	0.02 to 0.05%

Characteristics: see table above.

As the drilling mud is submitted to high temperatures within the hole, it is necessary to cool it. A cooling tower must therefore be installed. The efficiency of the tower depends upon its height and upon the temperature and humidity of the atmosphere. The difference in temperature before and after showering is 10 to 15 °C. This can be extended to 20 to 25 °C by installing a fan.

4.6 PREVENTION OF CIRCULATION LOSSES

This is of primary importance when drilling a geothermal well, and the counter-measures taken can be permanent or temporary. The permanent measure is to seal off the level at which losses occur, by cementing. Particular care must be taken in sealing off losses which occur once the casing has been put down and which were not apparent when actual drilling was in progress. The specific gravity of drilling mud is 1.1 to 1.2, whilst that of cement slurry is 1.6 to 1.8. Certain weak formations which stand up to the static pressure of the mud will break down under the static pressure of the cement slurry. The slurry should be allowed to rise from the level of the casing shoe to the ground level in a single stage. A pressure test should therefore be carried out every 50 m when drilling, by sealing the blow-out preventers, to determine the behaviour of the formation in regard to the difference in static pressure between mud and cement. If a breakdown occurs, the cement slurry should be injected into the formation.

Temporary measures consist of plugging the formations with nutshell powder, sawdust, cotton seeds, celophane, fibre scrap, etc. These vegetable materials are carbonised in the course of time once the well has been completed, and the steam production capacity of the formation is restored.

Holes are sometimes drilled without any plugging against lost circulation. This practice is not recommended for steam production because drill cuttings tend to be driven into the formation and to clog its natural voids.

4.7 CEMENTING

When cementing a geothermal steam well, the most important factor is that the cement slurry should rise uniformly and continuously from the casing shoe to the ground level. For this to be achieved, the filter cakes of the drilling mud should never be thick, even at high temperature. Also sufficient clearance must be assured between the casing and the wall of the hole for the void to be uniformly filled with cement. The casing must therefore be correctly centred by the use of centralisers. If the mud cake is too thick, it may be necessary to use 'scratchers'.

The cement slurry should be prepared in sufficient quantity, generally 1.5 to 1.7 times the estimated amount required, and injected by a large capacity pump which will prevent it chanelling.

If the slurry has not reached ground level when the cementing operation has been completed, the annular section between the double casings at the head of the well should be immediately sealed off and an attempt should by made to inject cement through the annular section in the reverse direction. If this is not possible, tubing can be lowered into the annular section and cement injected through it. When this procedure is followed, a pocket of water may remain trapped between the cement injected from below and that injected from above. When this happens with the intermediate casing, the pocket can later be explosively expanded when the steam begins to flow, and the casing may collapse inwards.

If therefore it has not been possible to fill with cement the annular void to ground level, it is generally preferable simply to provide a rigid support to the head section of the casing and not to attempt to inject cement from the surface. If the level of cement outside the casing seems to be very low, the hole may be re-cemented by means of a gun perforator.

If it is foreseen that the formation cannot be adequately cemented whilst drilling, or that it will be difficult to cement the whole height of the hole in one stage because the level to be sealed off is too deep, it is best to plan for two-stage cementation from the start.

4.8 CEMENT

Ordinary Portland cement is adequate for geothermal steam wells up to temperatures of 150 °C, but is insufficient above that. In general, the strength of hardened cement decreases, and its brittleness and permeability increase with time, if the temperature during the period of hardening rises above 120 °C. This can be prevented by mixing with silica flour. Cement to A.P.I. standards is classified into eight categories from 'A' to 'H'. These are used in oil wells according to depth, temperature, pressure, sulphate resistance, etc. For use in geothermal steam wells these cements must be supplemented with 30 to 50% of silica flour to reduce brittleness and permeability when hardening.

A method of decreasing the specific gravity and improving the adiabatic properties of cement is to mix perlite with the slurry. In countries where silica flour is not readily available, fly ash from thermal power plants can be used instead. Fly ash contains 55 to 60% silica, and a mix of 50% Portland cement with 50% fly ash (pozzolana) is used at a 50% water/solid ratio.

The addition of 0.3 to 0.5% lignin-sulphate derivative or polyhydroxy-carbonate may be used as a hardening retarder, whilst 0.4 to 0.6% methylol naphtalene sulphonic acid may be added as a dispersant, for thickening times of 2 to 4 hours, as required.

Gases such as H_2S, H_2SO_3, etc. are mixed with the steam and hot water in geothermal wells. They can produce heavy corrosion of the casing and can also have a harmful effect on the cement. It may therefore be necessary to use acid-resisting cement where geothermal steam is accompanied by low pH hot water.

4.9 DIRECTIONAL DRILLING

Well spacing for geothermal wells of 500 m to 2,000 m depth is usually 100 to 300 m. Directional drilling can therefore be achieved fairly easily even in hard rock formations. If therefore there are difficulties in siting a well such as those of access, delays in negotiation with the land-owner, or delays in obtaining the approval of authorities, it may be preferable to adopt directional drilling.

The direction and angle of a hole cannot be measured at high temperature. The hole must therefore be diverted at shallow depth when the temperature does not constitute an impediment. For instance, once the intermediate casing has been put in place, directional drilling can be started near the surface, and the hole deflected by one degree in the required direction for every 10 m of advance. The hole will thus have been deflected by 25 to 30 degrees by the time the level for the production casing has been reached. The production casing can then be placed and the hole continued in a straight line without further deviation. In hard formations turbo drill equipment is used as a deflecting tool.

4.10 DRILLING FAILURES

Particular failures which may occur while drilling geothermal wells are as follows:
— sticking of the drilling column because of steam or gas blow-outs;
— sticking of the drilling column because of rapid loss of mud circulation;
— parting off of the tool joint on the drilling column when drilling at high temperature in the winter season;
— breaking off of the threaded section on the drill collar, when encountering hard formations, etc.

Failures upon completion of a well can be due to:
— pressure collapse of the production casing due to rapid temperature rises in water pockets possibly trapped in the cement between the intermediate and production casings, due to heating by the steam flow;
— parting off the casing couplings when the outside of the production casing is inadequately cemented;
— rupturing of the casing below the main valve at the wellhead by large quantities of rock and dust carried up by the steam.

COUNTER-MEASURES AGAINST DRILLING FAILURES

Prevention of steam and gas gushes. The reservoir formations of steam and hot water must always be cooled whilst drilling so that their temperature does not rise above boiling point. When the temperature of the circulating mud is abnormally high, the mud should be passed through a cooling tower. Particular attention must be paid, especially at shallow depths, to excessive temperature rises due to hot steam entering the hole through fissures in the steam formation. Once a steam blow-out has occurred, the preventers at the head of the rig must be closed. But if the casing carrying the preventers is not well grouted in a competent formation, or if the cementing is inadequate, steam may then begin to flow outside the casing and around the rig. The formation holding the casing which carries the preventers must therefore be carefully selected, and cementing must be carried out perfectly.

Once the preventers have been closed, the steam blowing formation must be cooled by pumping cold water into the well, and the gush of steam must be stopped as soon as possible. If there is delay in closing the preventers, this may become no longer possible because of deformation of the ram. It may then be necessary to await a collapse in the well, which will weaken the gush of steam.

Steam blow-outs can also occur whilst raising the drill-pipe, because of inadequate cooling of the mud or insufficiency of mud, or because of swabbing action caused by thick mud and clogged drill strings, especially near the collar. It is therefore essential to use good quality mud capable of withstanding high temperatures. Trouble can also be avoided if the drill strings are provided with a back valve or kelly cock. Without these, even if the preventers

are closed, steam can rise through the drill-pipe and rotary hose, and the latter will be split by the heat and pressure, thus causing a blow-out and damage.

Preventing the drill-pipe from sticking through rapid circulation losses. When drilling in the productive zone, circulation losses are desirable because they indicate the presence of steam layers and are a guide to the productibility of the formation. As a result, precautionary measures tend to be neglected. The more pronounced the circulation losses, the quicker will the mud tend to settle downwards, carrying with it washings off the wall of the hole. The drill-pipe must therefore be raised as soon as possible, or else it will be liable to stick. Care must also be take nnot to use an oversized drill collar at the depths where circulation losses are expected to occur.

Recovery of a stuck drill-pipe. A stuck drill-pipe may be recovered as follows:

(a) *by immersion in lubricating oil.* Heavy oil, or special oil of high viscosity, is delivered around the stuck section of the drill-pipe to loosen it. The position of the stuck section may be estimated from the elongation of the pipe when force is applied to pull it. Once loosened, by immersion in oil for 8 to 24 hours, the drill-pipe can generally be pulled clear. This method has often been used in oil wells, but so far has not given good results in geothermal drilling because of the high temperatures encountered;

(b) *by using bumper and rotary jars.* When the bore is comparatively clean, the stuck drill-pipe is unscrewed at the joint section as deep as possible. A bumper or double acting rotary jars are then assembled on the drill-pipe which has been equipped with drill collar. This is run down the hole and screwed into the top joint of the stuck drill-pipe. Up and down blows are then struck onto it by the jars. If the length of the stuck section of pipe is limited, the pipe can be successfully loosened;

(c) *by using back-off tools (provided the temperature in the drill-pipe can be maintained at 90 °C for around 30 minutes after cooling).* It is often difficult to unscrew the joint of a drill-pipe immediately above a stuck section. Back-off tools can then be used, previously giving a counterclockwise torque to the drill-pipe by means of the table and exploding a detonator within the joint to be unscrewed. The joint section to be unscrewed must be properly identified by a collar locator. The location of the stuck drill section can be approximately identified by the elongation of the pipe on pulling. A magnetic detector is sometimes used to locate the stuck section more accurately. Once the joint has been unscrewed, the drill-pipes above the stuck section may be withdrawn;

(d) *reaming out the stuck drill-pipe.* Once the drill pipe above the stuck section has been withdrawn, the latter must also be recovered. The outer periphery of the stuck pipe is reamed over a length of 30 to 40 m by a casing pipe carrying a reaming shoe. A left-hand steel tap attached to a drill-pipe is then lowered and screwed into the joint of the stuck pipe to unscrew it. Recovery of stuck pipe can best be carried out four lengths at a time, provided the joints to be unscrewed have previously been loosened by exploding a detonator. Once all the drill-pipe has been recovered, the drill collar is reamed over by a casing pipe, as for the drill-pipe. If no heavy-duty back-off tools are available, the hole is reamed down to just above the bit, taking care that the reaming pipe does not stick. Finally, the remaining drill collar carrying the bit is recovered by a tap equipped with rotary jars, which screws into the drill collar.

Parting off of a drill pipe joint. When a high temperature well is drilled in winter, the pipe joints expand under the high temperature. When cold water is fed into the drill-pipe, the male screw section of the joint is liable to shrink and to part from the drill pipe. If this passes unnoticed, and drilling continues, it can lead to a serious failure. Such a joint can easily be recovered by a 'fishing tap' or an 'overshot' if noticed immediately. If drilling is continued without noticing this mishap, the cuttings will sink down the hole and stick to the disconnected length of pipe. This may then require reaming out. To avoid this, special care must be taken to inject cold water or mud into the well and any weakened joint must be replaced.

Break-off of the drill collar. This can occur when drilling with a heavy bit load and slow penetration in hard formation. The thread of the drill collar is repeatedly subjected to strong shocks when drilling in hard formation and is liable to fatigue through chattering. The drill collar can easily be retrieved by means of a fishing tap, provided the failure has been noticed in time and drilling has not been continued. To avoid the drill collar breaking off, the cutters of the bit must be selected in relation to the hardness and other characteristics of the formation. Selection of the proper bit is important. Once the hole has been completed, all threaded sections of the drill collar should be inspected by colour check and all defective collars rejected.

Collapse of production casing due to water pockets. This may lead to a collapse of the production casing at depths above that at which the intermediate casing has been put down. The only means of recovering a collapsed production casing is to ream out the cement around the production casing with a well connected pipe of suitable size from the cellar of the hole, to screw a large left-hand casing tap into the head of the production casing at the reamed out section, to fish out the damaged casing step by step, and finally to screw a new casing into the lower joint in place of the damaged casing. If the production casing is severely collapsed, it is necessary to correct to some extent the inside diameter of the damaged casing previously with a casing roller swage. To avoid collapse of the production casing, it should be cemented into place in the manner described above (Sections 4.6 and 4.7).

Parted coupling on the production casing. This may occur because of frequent temperature changes, especially if the outside of the casing is not properly cemented in. If possible, the disconnected upper part of the casing is lifted out and a guide of soft metal, which can be easily cut, is attached at its lower end. This carries a protrusion which can clamp inside the disconnected coupling and reconnect the casing. Once the casing is reconnected, the guide and bit are cut out. If the upper part of the casing cannot be recovered, a casing of smaller diameter is lowered to line the hole from top to bottom.

Rupture of casing below the main valve at the wellhead due to abrasion by rock dust. When a well has been completed and begins to blow steam, a large amount of rock dust can sometimes be carried by the steam for many hours. This can have a sandblasting action and wear through the steel pipe at the wellhead. Such an aperture grows rapidly and must be closed as soon as possible. This can be done by inserting a smaller casing with packing attached to its end, through the fully opened main valve, to a point sufficiently below the rupture. Light weight lifting tackle is used. The rock dust then passes through this inner casing, and the permanent casing can be repaired by welding a patch on the outside. (Example of success: Matsukawa, Japan.)

5. Air drilling

Ordinary drilling is by the rotary method with circulation mud: air drilling is rotary with circulating air instead of mud. This method has been tested over several years in various countries. Particularly good results have been achieved at the Geysers, California.

Particular features of this method of drilling are:
(i) high drilling speeds and low drilling costs. (Speeds 3 to 4 times greater, bit life 2 to 4 times longer, than with mud drilling);
(ii) no damage to the production zone from circulating mud injected during drilling.

Air drilling has, however, certain disadvantages, and is unsuitable for formations bearing excessive water or with a strong tendency to slough. Mud drilling must then be used.

It is the usual practice first to drill the formation with mud and then to resort to air drilling; or to drill with air in the production zone once the production casing has been put down and cemented in.

In many cases, once the production zone has been drilled, the hole is completed without inserting a slotted liner. If a liner is required, the hole is first cooled with water, which is then replaced with mud, and the slotted liner is inserted. The same practice is followed when the production casing is inserted and cemented after air drilling.

The air drilling rig is basically the same as for ordinary rotary drilling with mud circulation. It is adapted for air/mud circulation. The main parts of the rig are shown in Figure 4.

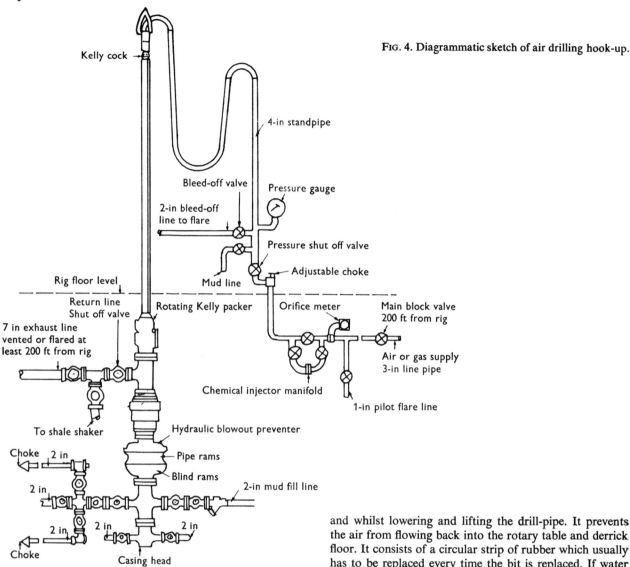

FIG. 4. Diagrammatic sketch of air drilling hook-up.

Kelly cock

4-in standpipe

Bleed-off valve

Pressure gauge

2-in bleed-off line to flare

Pressure shut off valve

Rig floor level

Adjustable choke

Return line Shut off valve

Mud line

Rotating Kelly packer

Orifice meter

Main block valve 200 ft from rig

7 in exhaust line vented or flared at least 200 ft from rig

Air or gas supply 3-in line pipe

Chemical injector manifold

1-in pilot flare line

To shale shaker

Hydraulic blowout preventer

Choke 2 in

Pipe rams

Blind rams

2 in

2-in mud fill line

2 in   2 in   2 in

Choke

Casing head

5.1 MECHANICAL EQUIPMENT (ATTACHMENTS TO THE ROTARY DRILLING RIG)

Primary and booster compressors. In air drilling, an air compressor and booster compressor replace the pump used for rotary drilling with mud. The primary compressor must provide an air velocity of 2,000 to 5,000 ft/min in the annular section between the hole and the drill-pipe. The booster compressor must provide sufficient pressure to blow out the water accumulating in the hole. The pressure will depend on the drilling depth foreseen. The capacities of the primary and booster compressors at the Geysers, California, for depths of 1,500 to 2,000 m were respectively:

700 ft³/min × 300 psi × 5 units (one spare)
1,500 ft³/min × 1,200 psi × 2 units.

Rotary drilling head (rotary Kelly packer). This is a preventer attached to the top of the double gate blow-out preventer. It carries an O.C.T. type ram which closes the space between the wellhead and the drill-pipe whilst drilling

and whilst lowering and lifting the drill-pipe. It prevents the air from flowing back into the rotary table and derrick floor. It consists of a circular strip of rubber which usually has to be replaced every time the bit is replaced. If water cooled, it can be used up to temperatures of 250 °C.

Inside blow-out preventer. Attached to the lower part of the drill-pipe this acts as a back valve to prevent steam from flowing out of the drill-pipe whilst lifting and lowering.

Drilling bit. The nozzle of a mud drilling bit can be enlarged and adapted for air drilling. Special bits for air drilling can also be purchased with specially adapted nozzles and air cooled bearings.

Discharge pipe. Air, dust and cuttings are blown through a discharge pipe, the area of which must be the same as, or smaller than, that of the annular space between the casing and the drill-pipe, to prevent cuttings from accumulating at the drill collar. Some discharge pipes are equipped with water sprays to ease the disposal of dust and cuttings.

5.2 COUNTER-MEASURES FOR WATER INTRUSION

As water intrusion from the formation increases, the efficiency of air drilling drops, and it may ultimately become necessary to resort to classical drilling. The following counter-measures are described in three stages which differ according to the capacities and pressures of the primary and booster compressors available.

Very small water seepage (500 l/h or less). Even with this small quantity of seepage, return cuttings tend to settle and the drill-pipe may stick with gradual increase in pressure and torque. This can be corrected by injecting finely ground calcium stearate and silica gel through the pump into the air stream, in quantities equivalent to 1 to 4% of the weight of cuttings normally ejected from the hole. Calcium stearate acts as a coating which repels water from the cuttings, whilst silica gel has 'anti-balling' properties. Both decrease the torque on the drill-pipe.

Medium water seepage (500 l/h to 10 kl/h). For seepages of this order, a foaming agent such as lithium stearate is used. This is termed 'mist drilling'. The foaming agent must be suitable for use in a steam well. Sundry agents are commercially available for use with fresh water, salt water, sulphurous water, etc. The foaming agent must be added to the water in the proportion of 3%, together with a corrosion inhibitor and a lubricant.

Large water seepage (over 10 kl/h). For seepages of this order the only remedy is to seal off the water by means of cement, plastic or other suitable material. These materials require the use of mud as a conveying agent. Silica tetrafluoride gas is sometimes used in the U.S.A. The water bearing level is sealed off by upper and lower packers and the gas is injected under pressure into the sealed off area. On contact with water, the gas forms a precipitate which effectively blocks the pores of the formation. If these methods are unsuccessful, it is necessary to revert to aerated mud drilling or conventional mud drilling. The choice between the two will be governed by the capacities of the compressors.

5.3 OTHER OBSTRUCTIONS AND REMEDIES

Cementing test. As water seepage into the well constitutes a serious impediment to air drilling, water leakage through the cemented casings must be carefully watched for. Once a casing has been lowered and cemented, it is essential that the cementing be tested and that any water leaks be sealed off by injecting cement under pressure.

Crooked hole. In air drilling, the line of pipes is not supported by surrounding mud, and bit reactions are strong. The bit therefore tends to deviate and the hole to lose its alignment. To maintain a straight hole, large diameter collars are used behind the bit. If necessary, a point reamer is attached to direct the hole from above, and the rod line can be equipped with fluted stabilisers set at 30 ft and 60 ft above the bit. Bit loading must be periodically checked.

Collapse of hole wall. There is as yet no known preventive against possible collapse. If the ground appears unstable, mud drilling must be used.

6. Well surveys

A well must be surveyed during and after drilling. The suggested survey is set out in Table 2.

TABLE 2. Well survey

Investigation item	Surveying instrument	Purpose of investigation
Temperature measurement	Point survey: (thermometer, geothermograph). Multiple survey (Kuster); continuous survey (thermistor).	Study of the change in temperature of the well, distribution and condition of geothermal heat. Presumption of steam and hot water layers.
Electrical log		Geophysical determination of formations.
Flow measurement	Spinner.	Flow measurement of steam or geothermal water at required depth.
Inclination measurement	Totoko and Murata type.	Measurement of direction and inclination of borehole, and charting the position and track of the well.
Pressure measurement	Amerada (bourbon tube). Humble type (spring).	Measurement of the pressure in the well, and presumption of the production capacity.
Collection of bottom hole sample	Bottom hole sampler.	Collection of geothermal fluid sample at required depth.
Cement bond log	Application of sonic log.	Investigation of cementing work.

7. Well spacing

There exist numerous studies regarding the spacing of oil wells, but so far there is no established theory concerning the spacing of geothermal wells. Even when the steam comes from a reservoir lying under a good cap rock, it is so often fed through faults and fissures that the degree of interference between two wells cannot be forecast, but will depend upon the degree and direction of the faulting and fissuring. In many countries experience has led to an empirical well spacing of 100 to 300 m for depths of 500 to 2,000 m.

8. Safety installations and precautions

8.1 ASSURANCE OF ADEQUATE COOLING WATER

Steam gushing during drilling operations takes a long time to control by means of the preventer, owing to the high temperature. Water supply must be sufficient for a possible emergency.

8.2 INSPECTION OF THE BLOW-OUT PREVENTERS AND THE TRAINING OF THE CREW IN OPENING AND CLOSING THEM

Periodical checks should be carried out and all crew members trained in the operation of preventers without loss of time.

8.3 ADEQUATE CEMENTING OUTSIDE THE CASING

The blow-out preventer is carried by the casing. The space around the casing must therefore be perfectly sealed off with cement to avoid steam gushing out when the preventers are closed. The casing shoe must be firmly held in a competent formation and any void outside the casing must be completely and uniformly filled with cement.

8.4 GAS DETECTION AND GAS MASKS

The gas emitted by a well may contain highly poisonous hydrogen and arsenic sulphides which can produce giddiness and eye injury even in low concentrations, and which may be lethal in high concentrations. Gas masks and detectors must therefore always be kept ready.

8.5 SAFETY CABLE SLIDE

When steam and hot water start to gush while drilling is in progress, men working on the derrick may be unable to reach the ladder. An escape cable carrying a 'bosun's chair' or other man-carrying appliance must therefore be provided between the derrick working floor and the ground.

9. Well repairs

Casing repairs have been described in Section 4.10 above. Repairs to a geothermal well usually have to rely on experience. Except in the simplest cases they may take many days to complete.

When a continuous steam flow cannot be maintained through an oversize casing in a production well incompatible with the energy of the steam formation, a smaller casing, chosen to correspond with the available energy, can be inserted into the original production casing. As an actual example, continuous flow from a production well originally equipped with a 9 5/8″ casing was achieved by inserting a $4\frac{1}{2}''$ casing from the top to the bottom of the well (Fig. 5). In another case a shallow formation producing water of low pH value was met in the production zone and the casing was badly corroded. The slotted liner was cut off half-way and recovered (Fig. 6-B); a cement plug was put into place between the level producing low pH water and the level producing steam (Fig. 6-C); a casing was then put down and cemented into position to prevent the intrusion of low pH water into the well (Fig. 6-D); and finally the cement plug was drilled out. The well was thus successfully repaired (Fig. 6-E).

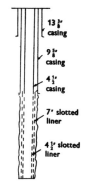

FIG. 5. An inner layer casing. A casing of $4\frac{1}{2}''$ diameter with slotted liner is inserted in the well to facilitate the outflow of steam. This smaller-diameter reduces to a minimum the temperature drop caused by adiabatic expansion of the steam and thus permits a continuous flow.

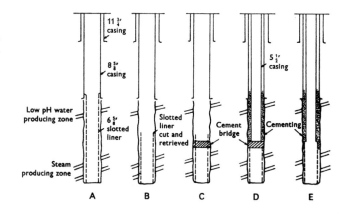

FIG. 6. Example of well repair.

Bibliography

Note. The reference 'Rome' relates to the United Nations Conference on New Sources of Energy, held in Rome in 1961 the proceedings of which were published in 1964 by the United Nations, New York.

The reference 'Pisa' relates to the United Nations Symposium on the Development and Utilisation of Geothermal Resources, held in Pisa in 1970.

BOLTON, R. S. 1961. Blow-out prevention and other aspects of safety in geothermal steam drilling. Rome G/43.

BRUNETTI, V.; MEZZETTI, E. 1970. On some troubles most frequently occurring in geothermal drilling. Pisa.

CIGNI, U. 1970. Machinery and equipment for endogenous fluid harnessing. Pisa.

——; GIOVANNONI, A.; LUSCHI, E.; VIDALI, M. 1970. Completion of producing geothermal wells. Pisa.

——; ——. 1970. Planning methods in geothermal drilling. Pisa.

CONTINI, R.; CIGNI, U. 1961. Air drilling in geothermal bores. Rome G/70.

CRAIG, S. B. 1961. Geothermal drilling practices at Wairakei, New Zealand. Rome G/14.

DENCH, N. 1970. Casing string design for geothermal wells. Pisa.

DURUCAN, E.; OLCENOGLU, K. 1970. Geothermal drilling and preliminary test operations at Kizildere, Turkey. Pisa.

FABBRI, F.; GIOVANNONI, A. 1970. Cements and cementations in geothermal well drilling. Pisa.

——; VIDALI, M. 1970. Drilling mud in geothermal wells. Pisa.

FISHER, W. M. 1961. Drilling equipment used at Wairakei geothermal power project, New Zealand. Rome G/49.

FOOKS, A. C. L. 1961. The development of casings for geothermal bore-holes at Wairakei, New Zealand. Rome G/16.

KARLSSON, T. 1961. Drilling for natural steam and hot water in Iceland. Rome G/36.

KATAGIRI, K. 1970. Effects of slotted liner casing in geothermal bores. Pisa.

MATSUO, K. 1970. Status quo of drilling and repairing of geothermal production wells in Japan. Pisa.

MINUCCI, G. 1961. La perforation 'rotary' pour les recherches d'énergie endogène. Rome G/66.

NIIJIMA, R. 1961. A study of the characteristics of rotary drilling practice in steam or hot spring wells in volcanic territory. Rome G/22.

NAKAJIMA, Y. 1970. Geothermal drilling in Matsukawa area. Pisa.

SMITH, F. W. 1959. Advancement in air drilling during 1958. *Drilling and production practice*, p. 244-248. New York, American Petroleum Institute.

SMITH, J. H. 1961. The organisation for and cost of drilling geothermal steam bores. Rome G/40.

——. 1961. Casing failures in geothermal bores at Wairakei. Rome G/44.

STILLWELL, W. 1970. Drilling practices and equipment in use at Wairakei. Pisa.

WOODS, D. I. 1961. Drilling mud in geothermal drilling. Rome G/21.

Well measurements

N. D. Dench

Ministry of Works (New Zealand)

1. Introduction

1.1 OUTLINE

Geothermal well fluid measurements are made for the following various purposes:
(a) Basic study of a natural resource;
(b) Assessment of an underground thermal reservoir for possible exploitation;
(c) Assistance in drilling operations;
(d) Appraisal of individual wells for production;
(e) Mechanical engineering design requirements, including safety of equipment and personnel;
(f) Legal requirements, for ownership, safety or waste disposal;
(g) Fluid sales;
(h) Plant operation.

If adequate data are gathered for (b) and (d) it is likely that most of the other needs will be satisfied.

The measurements described comprise:
(a) Reservoir investigation, in particular as to its size, permeability and temperature, and also the fluid composition and pressure;
(b) Well flow characteristics, specifically temperatures and pressures and the corresponding flow rates of the various constituents (steam, hot water and gas);
(c) Downhole engineering data, such as casing condition, mineral deposition, or levels of permeability;
(d) Miscellaneous observations carried out conveniently by the well measurements personnel.

1.2 PRACTICE

In that much care is needed in making and interpreting instrument observations, geothermal measurements differ little in principle from those in other fields of engineering. Frequently, high accuracy is needed; not so much for the observation itself as for the ability to show small differences over periods of time. Accuracy is dependent on the behaviour of the fluid system, the sensitivity of the method, the quality of the equipment and the standard of testing. Due care in the choice of method and equipment, and in training of staff, is necessary if the results are to be of value. It is important that detailed standard installations and procedures be adopted, and that when comparisons or trends are sought similar methods be adhered to. Frequent calibration, repeat tests and checks of equipment and methods are vital to sustained satisfactory results.

There is an unjustified tendency to accept as applying to all field measurements the degree of accuracy quoted by the manufacturer (or the standard reference) for the most favourable conditions. A further belief, often held unwisely, is that the accuracy of reading is necessarily an indication of the overall accuracy of measurement. A third fallacy is the assumption that downhole readings always reflect general conditions in the mass of the reservoir. In general, a critical attitude towards all results is desirable.

1.3 UNITS

The International (S.I.) system of units is primarily used in this article, but the equivalent former British units are also sometimes quoted, the conversion factors being listed in the Unit conversion table, page 11. Fluid pressures may be specified in 'newtons'/m² or in 'bars', but for many practical purposes, the bar, the standard atmosphere, and the kg/cm² may be taken as equal. Where megawatts (MW) appear in the graphs they represent total heat flow rather than electrical output.

1.4 STEAM

The properties of steam (and water) are known accurately, and are tabulated in standard steam tables, (see Bibliography). Geothermal well measurements make much use of these tables and of derivatives such as dryness and flash. Some of these inter-related properties are shown (Fig. 1 to 3) to illustrate the effects of pressure changes.

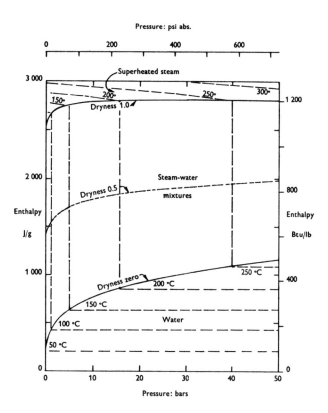

FIG. 1. Steam properties: enthalpy, temperature, pressure.

FIG. 2. Steam properties: dryness, pressure, enthalpy.

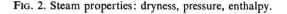

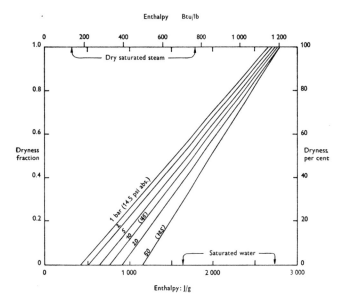

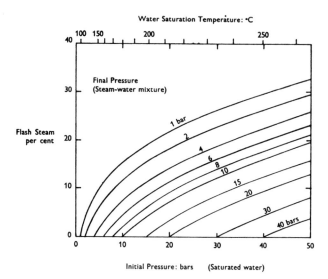

FIG. 3. Steam properties: flash steam, pressures.

2. Reservoir and well conditions

2.1 SUBSURFACE

It is seldom possible to make sub-surface measurements of the natural conditions, because:

(a) the cooling action of the well drilling operation lowers the local reservoir temperatures temporarily;

(b) the introduction of a hole in the formation permanently changes the mass permeability of the reservoir, and thus encourages new flow patterns.

In general, the larger the hole diameter and depth, the greater will be the resulting disturbance of the natural regime.

Normally, the well temperatures and pressures will be closest to reservoir conditions at the levels of greatest formation permeability. Throughout the section of drilled hole exposed to the reservoir (that is, uncased, or with perforated casing), measured pressures must be fairly close to local reservoir values, while temperatures may differ substantially if there is fluid flow up or down the hole between permeable levels. In that section of hole sealed from the reservoir by cemented casing, a change of phase can cause large pressure differences between well and formation, even while temperatures are similar. However, it is usual to place more reliance on the values measured below the solid casing depth.

As the accompanying examples show (Fig. 7), subsurface temperatures in water can vary rapidly and unpredictably with depth, and inversions are common, particularly in wells close to the boundary of the field. It is therefore clear that interpolation of temperature should be done cautiously, and extrapolation to greater depths, never. On the other hand, a pressure inversion cannot occur, and the pressure gradient does not change suddenly unless there is a change of phase (Fig. 6). Both interpolation and extrapolation in water, are often justifiable for pressures.

The gradient of pressure with depth is a direct measure of fluid density. If the temperature is also known, reference to steam tables will show what fluid type is present. As the common gases found in geothermal wells, like CO_2 and H_2S, are considerably heavier than steam at the same temperature and pressure, a fluid of density between those of steam and water is either:

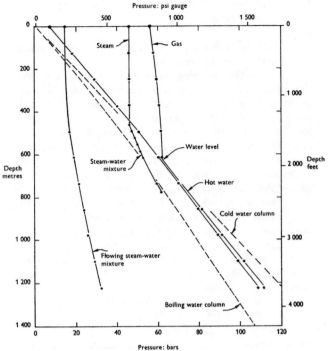

FIG. 6. Pressure in typical wells.

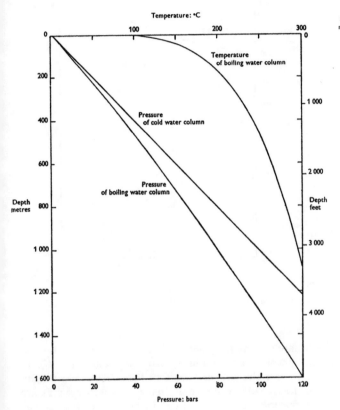

FIG. 4. Water columns: theoretical temperatures, pressures.

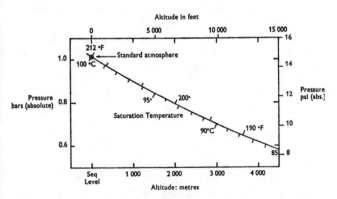

FIG. 5. Boiling conditions: variation with altitude.

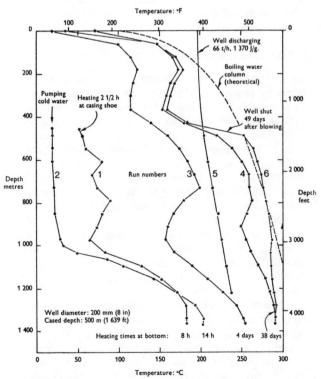

FIG. 7. Downhole temperatures at various well conditions.

(a) a steam-water mixture if at saturation temperature/pressure conditions;

or

(b) a gas (plus water vapour) if the temperature is below saturation for the pressure.

During the planning of a well-measurement programme in an unexplored field, when equipment is being arranged, an estimate must be made of the expected maximum values of depth, temperature and pressure. The capacity of the drilling rig is one guide for the depth. In hot water reservoirs the bottom hole pressure is unlikely to exceed the pressure due to a cold water column from the surface, and in dry steam conditions the fluid pressure will certainly not exceed that due to rock weight. Sometimes chemical analysis of surface springs can suggest maximum subsurface temperatures; and unless superheated steam is suspected, the value will be limited by saturation conditions corresponding to the pressure assumed. Convenient reference lines for comparing, or very roughly predicting, subsurface temperatures and pressures are the theoretical boiling water curves shown in Figure 4.

The timing of downhole measurements should be determined in relation to the disturbance caused by drilling or discharging the well. Depending on the permeability of the formations and on the amount of the disturbance, stabilizing of downhole conditions may be virtually instantaneous, or, at the other extreme, may take many months to complete. For any one well or depth, the recovery rates for temperature and for pressure are likely to be quite different. Some measurements detect slow recovery (Fig. 7), while others are designed to observe long term trends of stable values.

2.2 FLOWING CONDITIONS

Geothermal fluids—water, steam and various gases—may occur by themselves or in mixtures, and with impurities such as salts or sand. Some of the minor constituents have substantial commercial value, rivalling that of the heat, while others are only a nuisance to utilization or measurement. Flow measurements employ both physical and chemical means to identify and rate all of the discharge.

The fluid reaching the surface has approximately the same energy as that entering the well from the formation. Water below roughly 100 °C, and dry steam, reach the surface with little change; but water entering at higher temperatures, or steam-water mixtures, boil on the way up the casing, losing temperature and pressure in accordance with saturation conditions. The enthalpy and pressure of the mixtures govern the proportion of steam and the velocity of flow at any point.

The flow rate in a well is limited either by the permeability of the formations, or by the physical restrictions of the wellhead, where velocities are highest and may reach the critical value for the fluid. A well which taps zones of equal temperature has an output of practically constant enthalpy over the range of flow rates. These change according to the degree of restriction, while the dryness follows

the relationship shown in Figure 2. Figure 9 illustrates this sort of production.

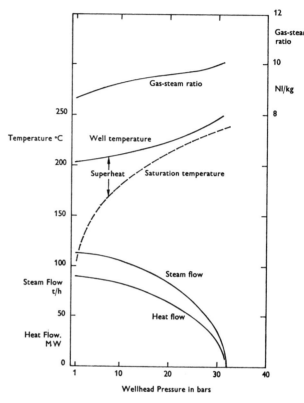

FIG. 8. Well flow: superheated steam and gas.

When a well flow is choked to the limit, it reaches a pressure maximum, either on complete closure, or, in some cases, while there is still a small output (or 'bleed'), in which case the wellhead pressure then decreases until full closure is made. Although the maximum discharge pressure is influenced partly by the rate of closure, it can be used as a rough indicator of trends. The lower pressure limit of a well's operating range is governed by reservoir permeability or surface pipework, as mentioned above.

If the well is producing from two or more horizons at different temperatures, an alteration of the wellhead throttle causes pressure changes in the flow which are likely to affect the proportions of fluid being drawn from each level. The output then has an enthalpy which varies with the wellhead conditions, even to the extent of a change of phase (Fig. 10).

A steady output is typical of most flows of dry steam or water. However, some flows containing water exhibit geysering, which appears at the surface as cyclic variations in flow-rate, pressure, and quality (temperature, or dryness). Most two-phase flows have small, rapid pulsations caused by the continuous boiling which takes place as the fluid rising in the well loses pressure. Initial boiling may occur in the formation, in the casing, or in the surface pipework; and as it may be associated with chemical deposition

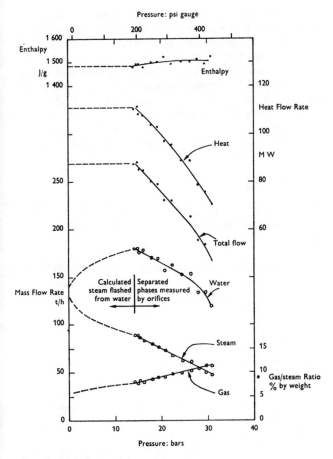

Fig. 9. Well flow: steam, water, gas.

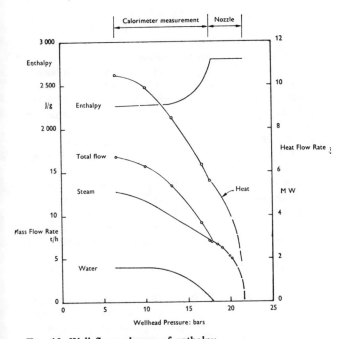

Fig. 10. Well flow: change of enthalpy.

of salts, testing for its position and effect is carried out. Another feature of two-phase pipe flow is that much of the water flows relatively slowly along the pipe walls, while the steam and remaining entrained water travel many times faster in the more central part of the cross section. The slow nature of some of the long term trends due to changes in the reservoir affects the accuracy of measurement necessary for their recognition within a reasonable time. Other characteristics of flowing wells which must be considered when planning measurements are:

(a) the enormous amount of energy released, which must be controlled safely;

(b) the very high noise level, especially with the drier discharges;

(c) the possible solids content, which can erode the pipework at bends, and which may need to be measured along with the fluid flows; and

(d) leakage of heavy, poisonous H_2S gas, which may collect in low enclosures like drilling cellars.

3. Pressure measurements

3.1 SURFACE GAUGES

The instruments used for measuring geothermal fluid pressures in surface pipework are the same as in conventional steam practice. Bourdon type gauges cover the whole range of pressures met, from sub-atmospheric to values exceeding 50 bars (700 psi). These instruments measure 'gauge' pressure, or the amount by which the fluid pressure exceeds the atmospheric pressure. While the day-to-day variation of atmospheric pressure is not significant for most well measurements, the average value for a particular field depends on its altitude (Fig. 5). Steam properties, as derived from the steam tables, are expressed in terms of 'absolute' pressures.

Atmospheric pressure is commonly measured by an aneroid barometer, or, more accurately, from the corrected height of a mercury column in an evacuated tube.

An accurate means of measuring low pressures, or pressure differences (such as across a flow orifice) is with a glass manometer or U-tube, usually containing mercury but sometimes water or other light fluid for very small pressure differences.

Pressure values may be indicated on circular dials, and/or recorded continuously on charts. The various pressure recorders may take disc or strip type charts, be spring or electrically driven, and operate over a wide range of speeds.

Although pressure gauges are basically uncomplicated instruments, proper care should be taken with their installation, calibration, observation and maintenance, as for all instrumentation. For instance, a correction should be made, if necessary, for hydrostatic pressure due to a column of liquid in the tube between the gauge and the point of measurement. Also, particularly in geothermal applications, bourdon gauges should be protected from high temperature and from rapid pressure fluctuations. For corrosion precautions, see article by Marshall and Braithwaite in this Review.

3.2 DOWNHOLE PRESSURES

The instrument used commonly for downhole pressure measurement was developed initially for petroleum wells, and subsequently modified in the clockwork chart drive mechanism to withstand the high temperatures of geothermal wells. It is a vapour-filled helical bourdon tube joined to a bellows unit, on the outside of which the well pressure acts. The bourdon tube, thus sealed off from the well, scribes a line on a cylindrically wound chart driven axially at a constant rate.

The gauge, approximately 30 mm in diameter by 2 m long, is suspended in the well by a 2 mm stainless steel wire through a steam gland at the surface. The wire is moved by a small winch with hydraulic braking and a depth meter riding on the wire. As the gauge takes several minutes to stabilize at any pressure, it is necessary to stop at each depth (e.g. every 100 m) for a satisfactory reading. Surface observations are made of time, depth, and wellhead pressure for a calibration check. To ensure flexure of the bourdon tube, the gauge is first lowered to well bottom, and measurements are then made on the way up.

When, later, the chart is removed from the instrument it is viewed in an optical, micrometer reader. Axial distances indicate time, which relate to depths, and offsets are a function of pressure, read from calibration curves for the approximate well temperature (as the instrument is not completely compensated for temperature). Corrections to the observed depths are necessary to allow for thermal expansion of the wire. It is clear that corresponding well temperatures must be known (Hunt, 1964; Wainwright, 1970).

Downhole pressures have also been measured by slowly forcing gas (e.g. nitrogen) down small diameter tubing, and reading the stable gas pressure at the surface. The tubing must be shifted for each depth to be measured, and again temperatures must be known in order to calculate the weight of the gas in the tubing (and to correct the tubing depth). This method is too cumbersome for general use.

With the advent of electric logging cables which will operate in the hot well conditions, pressure transducers with surface recording can be expected as instruments for the future. Simultaneous recording of downhole temperatures and pressures should be the objective.

When a well is known to contain only one phase, measurement of the wellhead pressure or of the water level, plus downhole temperature readings to give densities, allow pressures to be calculated throughout the depth of the well. Providing that the temperatures have been proved constant with time, the observation of water level or wellhead pressure alone can be a more accurate indicator of downhole pressure changes than logging with a pressure

gauge. The influence of varying atmospheric pressure on water levels is measurable, but corrections are required only for very accurate work. In areas near the sea, the effect of ocean tides may be greater.

The water level in a well is normally measured with a cylindrical float suspended on a wire line. Twin core cable can also be used to complete an electrical circuit when the water is reached. Some wells have a mixed phase zone between the steam and water columns, which prevents the detection of a recognizable water surface (Fig. 6, short curve).

3.3 CALIBRATION

Both surface and downhole instruments are calibrated by conventional dead-weight testers. For downhole instruments, a high-temperature, high-pressure vessel is necessary to hold the gauges at well conditions during calibration. Safe design and accurate control of temperature and pressure at the gauge itself, are essential. The vessel may be used also for calibrating downhole temperature instruments.

3.4 PRESSURE LOGGING

A convenient method of measuring well permeability is to conduct water injection tests after the completion of drilling. Water is pumped into the well at several flow rates, while downhole pressures are recorded at a fixed depth. This depth is not critical, but preferably should be close to the permeable level to avoid friction losses above, and density changes below. In the absence of precise information, the casing shoe level may be chosen. When pressure change is plotted against injection flow rate, the slopes of the lines provide an immediate guide as to relative permeabilities, and if the drilled hole is likely to be productive from isolated levels only, it is of doubtful advantage to try to convert the figures to absolute terms.

It is advisable to log the well for pressure distribution after the completion of drilling, and again when temperature stability has been reached. Intermediate runs are also helpful. Changing temperatures are accompanied by changing densities, and it follows that the pressure line slope will also alter. If the changes are great, and the pressures controlled by only one permeable level, that level may be identifiable by the pivot point of the pressure lines.

Figure 6 shows examples of pressure runs in various well fluids, including one of a steam-water mixture during well production. Such runs should be done with a weighted instrument, and a line tension indicator at the surface, to detect any likelihood of the instrument being blown out of the well.

When a well is shut after a period of production, immediate and subsequent pressure runs show the degree of drawdown in the reservoir near the well, and also the rate of pressure recovery. When studying the pressure trends with time, the reservoir engineer chooses a depth close to the production level (which controls the well pressure), and if two phases are present, generally compares figures from below water level.

For greatest value, a logging record should include full physical details of the well and its recent operating history, in addition to all observations, the depth corrections for wire expansion, and the pressures from the calibration. The presentation of the results is usually enhanced by including a graph, frequently on the same sheet as previous logs of that well. The measured values should be plotted clearly by points and joined by smooth curves, except that where a water level has been measured by a float, or inferred from the results, a sudden change in slope should be drawn. See the examples.

4. Temperature measurements

4.1 EXPANSION THERMOMETERS

Of the instruments which depend on the thermal expansion of a liquid, the mercury-in-glass thermometer has the widest use in geothermal work. It covers the full range of temperatures met, is stable and accurate, simple and cheap, and can be used in nearly all surface applications. Although other instruments give much superior results downhole, its maximum recording version can give limited but worthwhile information in subsurface work.

Liquid expansion instruments also include those in which pressure changes cause movement of a bourdon tube, as in pressure gauges.

Bi-metallic instruments are commercially available for surface applications, and their principle has also been applied in the 'geothermograph', (developed specially for use in geothermal wells), in which the swing of a bi-metallic reed is scratched across a plate whose longitudinal movement is controlled by jerking the wire line at each depth.

The geothermograph may be run, read and calibrated with the same surface equipment as used for the downhole bourdon-type pressure (and temperature) gauge. The highest temperatures recorded in geothermal wells, just over 300 °C, have been measured with this instrument. Its accuracy is not better than 2% of the range.

The pressures of various vapours are also employed in bourdon type instruments to indicate temperature change. The vapour pressure instrument relies only on the temperature of the thermometer liquid providing the vapour, and is successful in geothermal measurements. Indeed the same chart and drive mechanism are used alternatively with either temperature or pressure sensing bourdon elements. A feature inherent in the vapour pressure thermometer is the close scale at low temperatures, and higher accuracy at high temperatures, a disadvantage partly countered by using two limited range elements (Hunt, 1964; Wainwright, 1970).

4.2 THERMOCOUPLES

Use has been made of thermo-couples both for surface and downhole measurements. The references give details of the life, range and accuracy of various metal pairs, and illustrate the several wiring choices possible. The greatest difficulty in downhole use is in finding a thermocouple cable which will retain its insulation for an economic number of runs. A good surface application is in the controller circuit of the calibrator.

Surface indication is given of downhole conditions, either by measuring the direct deflection on a millivoltmeter, or more accurately by the null method, in which an electrical balance is measured with a potentiometer. Either direct indicating, or automatic recording, may be used in these measurements (American National Standard, 1964).

4.3 RESISTANCE THERMOMETERS

These sensors take advantage of electrical resistance changes which accompany temperature variation. Dependable materials include:
(a) platinum wire, which has a positive increase in resistance with temperature rise;
(b) metal oxide beads, with a larger, negative characteristic.

These latter are called 'thermistors'. The references show a number of circuit systems which may be used, as well as an array of measuring instrument hoop-ups possible, as for thermocouple practice. A 3-wire cable and a null balance circuit is suitable for downhole work.

Both thermocouples and resistance thermometers can be instrumented to provide a direct digital output suitable for electronic data storage or computation. Resistance thermometers are considered more accurate, and for downhole work operate with normal conductor cable. Both of these thermometer types approach the ideal of continuous recording with little lag, in contrast to the set-wise operation necessary with the geothermograph and the vapour pressure thermometer. The cable-run detectors require more elaborate winch gear and instrumentation than do the wire run thermometers, and are correspondingly more costly.

4.4 TEMPERATURE LOGGING

One of the first temperature runs in a newly-drilled well may be done to detect the permeable level. After injecting water at the surface for several hours, while allowing the bottom section of the hole to heat up, a sharp rise in temperature indicates the water loss point, and hence the highest level of good permeability. More conclusive results can be achieved at times by allowing an initial heating period, while a change to a higher pumping rate, and/or successive temperature runs, may show up a second, lower permeable depth. Unless volumetric calculations are made, incorrect assumptions of downhole happenings are likely. Figure 7 shows an effective water loss run amongst a series of heating runs.

It is useful to take several temperature logs between drilling completion and the reaching of stable temperatures, a condition which may take several hours, weeks or months. Although there is a natural eagerness to blow a well as it comes under pressure, reservoir conditions should be recorded after as little disturbance as possible, and it sometimes happens that different stable temperature patterns are evident before and after initial discharge. Predictions of stable conditions from early heating runs cannot be made with confidence.

Temperature runs are sometimes made during well discharge, but pressure readings are then generally adequate. However, after well closure, runs may be required to measure the rate of temperature recovery.

Water temperatures in wells of 100 mm (4 in) diameter and smaller are generally considered to reflect local reservoir conditions fairly accurately, but those measured in the gaseous phase, or in wells of 150 mm (6 in) or larger, should be treated with caution. However, the values at local peaks and lows, particularly at levels of high permeability, are probably close to reservoir temperatures. The movement of fluid in the well accounts for these effects.

Just as for pressure logging, full physical and operating details of the well should be recorded, in addition to the measurement data. Again, if a continuous record was not made, the actual measured points should be plotted, but, unlike the graphing of pressure, the points should be joined by straight lines in most cases.

5. Flow measurements

5.1 TEST PROGRAMME

Both the drilling record and the subsequent measurements in the closed well give a rough indication of the output to be expected, and therefore of the method of flow measurement likely to be the most suitable. The drilling log should note the temperatures and losses of drilling fluid, in addition to the occurrence of gas and wellhead pressures.

If the completed well stabilizes with a pressure at the wellhead, it will of course discharge when opened, although sometimes the flow may become intermittent or even cease altogether. On the other hand, the lack of a wellhead pressure does not necessarily prohibit discharge, if for instance a relatively cold column of water can be lifted off, or depressed and heated before releasing. A comparison of downhole pressures and temperatures with saturation conditions helps in the planning of an artificially induced blow

For wells producing steam, a vertical blowpipe causes the least interference to the flow, and, if sand is discharged, limits erosion to the high-velocity region beyond the valves. If a similar wellhead is installed on each well at its first blow the reading of wellhead pressure, plus visual observation, allow rough estimation of the discharge rate (relative to wells measured earlier).

The observation of wellhead pressure alone, over a period of time, provides a useful indication of any changes in either quantity or quality of flow. Frequently, the first few hours or days of discharge see substantial changes in flow, either sudden in the case of the output of dirt, or gradual when reservoir pressures are adjusting to production conditions. Commonly, the well measurement programme will comprise a full range of output testing at intervals of several months, linked together in time by wellhead pressure readings, preferably automatically recorded.

A complete well output test will report at several wellhead pressures:
(a) The flow rates of the total mass;
(b) The quality of flow (enthalpy or dryness) or the temperature of single phases;
(c) The chemical constituents of the phases;
    Also:
(d) The highest and lowest pressures attainable;
(e) A description of the test itself;
(f) A brief history of the well;
(g) Reference to associated observations, e.g. downhole pressures during discharge, or the behaviour of nearby wells.

Simpler tests will sometimes be adequate; e.g. the measurement of a single phase at one pressure.

The various test pressures are achieved by throttling the flow with valves or different diameter orifices. Except when the flow at one specific pressure is needed, it is often more convenient to choose the choke setting or device, and to test at whatever pressure results. Graphical interpolation for other pressures can then be made, if the well has a smooth relationship between output and choke size.

When a well has an unstable output at any one throttling, its behaviour in the region of its expected production pressure must be studied with particular care, as described later.

Like other measurements, the method used for flow testing will be governed by the well characteristics, the resources available, and the accuracy needed. Particularly when new techniques are being introduced, it is wise to make repeat measurements by the same method, and check measurements by other methods, to ensure that no precautions are being overlooked. Also, the prior operation of a well sometimes affects the output, so that testing with both increasing and decreasing stages of throttling may be necessary. When changes of output with time are being studied, the differences are most reliable where the same method is used for each test.

5.2 ORIFICE MEASUREMENTS

The measurement of single phase flow under pressure is best made by observing the pressure difference across a circular restriction in a straight run of the flow pipe. While a venturi causes the least net pressure loss, most installations use the less costly nozzle or sharp edged orifice. The pressure difference, measured by a manometer, is dependent on both the properties of the fluid and the physical dimensions of the meter. Orifice geometry, the flow calculations and the errors involved, are described fully in the standards of engineering and national institutions, and in the ISO publication R541-1967 (E). As the measurement is upset by a change of state, saturated water must be cooled or pressurized (hydrostatically by lowering the orifice pipe) to avoid boiling at the orifice. Meters which convert manometer differential pressures into dial readings or recordings on charts are made commercially, and will be used in permanent installations where the range of flows is known. Early testing is done better with a series of different sized orifices and direct reading manometer.

For well flows of superheated steam or of hot water, testing should include temperature measurement, to enable the heat flow to be calculated (Fig. 8). With saturated mixtures, the pressure reading is sufficient.

Where gas is present in the flow, it is measured conveniently by sampling the flow and analyzing chemically. The orifice calculations must be worked on the basis of the mixture of gas and steam. If its flow rate is small, the total steam output may be condensed by cold water injection upstream of the separator, and only the gas measured through the steam line orifice.

For orifice testing, steam-water mixtures must first be separated and the two phases measured individually.

The combination of cyclone separator, with sharp-edged orifices and manometers for the water and steam fractions, is the standard method of measurement for two phased mixtures, and is the reference against which other methods are calibrated. Calculations of enthalpy and heat follow from the flow rates and pressure.

5.3 CALORIMETERS

Whereas orifice-type meters measure actual flow rates, calorimeters measure the volume and heat produced during a convenient time interval. Essentially, the calorimeter is a tank partly filled with cool water into which the flow is passed for a measured period. From the gains in volume and temperature of the water can be calculated the heat and flow rates.

Accurate time duration is achieved by using a quick-acting swinging arm discharging the flow. While it is barely practical to cater for very large outputs, flows of up to 100 t/h have been measured in a tank of about 15 m³. Easily portable smaller tanks are useful for low output wells discharging steam-water mixtures.

Large well flows may be measured cheaply by traversing a 10 mm diameter sampling tube across the flow and leading the sample to a small (200 l) calorimeter. The timing of the sampler at each radius must be proportioned to the area of flow it represents, and the whole unit must be calibrated regularly against a separator/orifice setup.

While, in theory, all flows may be measured by such a sampling calorimeter, superheated steam and hot water are usually best measured in other ways, and calorimetry is confined largely to mixed flows. The tanks, however, may

be used also for volumetric measurement such as the calibration of hot water weirs and flumes.

5.4 MISCELLANEOUS

For hot water flow measurement in the open, the venturi flumes and sharp-edged weirs developed for cold water use are suitable, but, depending on the method of head measurement, some calibration or adjustment of the formulae constants may be necessary. Also, deposition of salts can change the physical dimensions, and water vapour may obscure surface levels. A gas purge and pressure gauge device can be used to overcome the latter difficulty. A thermometer is needed if the heat flow is to be recorded.

When the hot water can be measured with tolerable accuracy, as in the weir of a separator-type well silencer, measurement of the critical lip pressure of a steam-water mixture enables the enthalpy and flows to be calculated (James, 1962). The method can be calibrated against separator/orifice measurements, and for silenced wells is a cheap means of recording output continuously.

If sand is present in fluid output, an elementary measurement of the percentage of solids, by conventional sediment sampling, will be useful.

5.5 FLOW MEASUREMENT RESULTS

The flow properties observed directly are often not those of most use, and calculations must be made from the corrected readings and the steam tables (or derived tables) to find those needed. For instance, from calorimeter records of mass and heat flow enthalpy and dryness fraction must be calculated in order to know the steam flow.

As the results at specific pressures are not necessarily the most suitable, and in any case do not fully represent the well characteristics, it is usual to plot the observed values against wellhead pressure, and to draw smooth curves where the well behaviour justifies it. To indicate the variation in results it can be useful to plot also the calculated points for the derived parameters. Figures 8 to 11 show examples of various types of wells, measurement methods, and presentation.

When a well shows cyclic variations at any or all pressures, the test method and times must be selected to demonstrate its operational characteristics. If the well is to be connected to a constant pressure system, testing it at fixed throttling does not necessarily help in predicting its production performance. As orifice measurements are virtually instantaneous, and calorimeters are averaging devices, they may well give different results.

Figure 11 shows a series of instantaneous results taken at 15 minute intervals on several days on a well having fixed throttling. In spite of the well's quite variable behaviour under this restraint (see upper plots) it would produce a steady flow of dry steam when fitted with a suitable separator and controlled to a constant, high steam line pressure (see lower plots). It is possible, sometimes, to vary the throttling to maintain a constant pressure for testing.

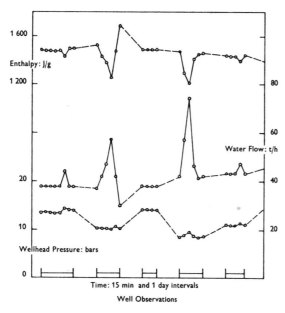

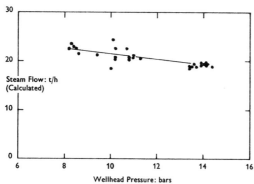

FIG. 11. Well flow: erratic discharge with constant choke.

In addition to the graph, a flow measurement record should contain the original observations and the basis of calculation for the derived figures. Section 5.1 above lists the types of information which should accompany the flow values. The inclusion of a wellhead pressure recorder chart (suitably annotated) ensures that dates, times and general well behaviour are not forgotten.

Some estimate of the likely accuracy of the results is desirable, but a valid one is not easy to obtain. In general the main methods can be listed in descending order of accuracy by: (a) orifice; (b) calorimeter; (c) sampling.

6. General

6.1 DOWNHOLE OPERATIONS

Many downhole instruments have been developed for use in the petroleum industry to provide information about the reservoir, or to help in the drilling. While most of these are of value in geothermal drilling, understandably only some are suitable for steam service. High temperatures have limited the type and range of equipment which can be used, and have confused interpretation of the results.

Wire line measurements include (Dench, 1962):
(a) deviation surveys, which record the amount and direction by which the drill pipe is off vertical;
(b) casing calipering, to check for corrosion or fractures;
(c) soundings, with various diameter cylinders, to establish the extent of blockages caused by casing deformation, chemical deposition, or sand entry.

Electric cable services during drilling include:
(a) hole calipering, for the calculation of casing cement quantities and possibly some production parameters;
(b) bond logging, to check casing cementing;
(c) casing collar location, for depth control of other operations;
(d) casing perforating, for fluid production.

Various geophysical logs, such as electrical resistivity, have been made in geothermal wells in order to correlate the formations, to indicate porosity, or to calibrate surface geophysical surveys (Japan Geothermal Energy Association, 1969). However, their economic justification is doubtful, and more value is likely to accrue from spinner logs which are being developed to show whatever flows occur in the wells when closed or discharging.

6.2 SURFACE OBSERVATIONS

Because of their availability and special skills, personnel who do the well measurements may be called upon to include in their duties other observations not directly concerned with geothermal wells, such as the following:
(a) the measurement of natural surface thermal activities —e.g. hot spring flows and geyser discharges;
(b) meteorological observations—e.g. temperatures, relative humidity, barometric pressure, rainfall, wind speed and direction, snowfall, etc.;
(c) estimation of surface and underground cold water resources;
(d) pollution observations of atmosphere, surface drainage systems, and groundwater.

6.3 ORGANIZATION

Whether the measurements are being done under contract or directly by the operating authority, practical and safety considerations will normally require a field staff of at least one technician and a helper. With competent personnel, few wells, and good equipment, two men may be able to handle all work including observation, instrument servicing, and reporting. In a larger field, where more people are necessary, there will be a tendency for staff to specialize in narrower sections of the work, but it is desirable for some to be trained in all phases. Previous experience in steam, hydrological or petroleum practice is useful.

Competent engineering supervision will ensure that the right observations are made in the right way and at the right time. Much effort can be wasted and much useful data lost, in the absence of close control. While it is desirable to establish standard measurement programmes, they should be reviewed from time to time for necessity and sufficiency.

Safety is an essential part of measuring geothermal wells. Providing realistic assumptions are made as to possible temperature and pressure conditions at the wellhead, adequate design of test equipment can be made as in normal mechanical engineering. Operating procedures for both the wells and the measurements should be issued in writing, not only to safeguard lives and property, but also to avoid unscheduled disturbances to well conditions and consequent delays while awaiting recovery. If H_2S or other poisonous gases occur, precautions should include good ventilation, detectors, and emergency breathing equipment. In other respects, the usual construction safety rules apply. Where noise is a problem, personnel should wear ear muffs.

Bibliography

ABBREVIATIONS

H.M.S.O.—Her Majesty's Stationery Office.
Pisa—United Nations Symposium on the development and utilization of geothermal energy, Pisa, 1970, the proceedings of which have been published by the Istituto Internazionale per le Ricerche Geotermiche, Lungarno Pacinotti 55, Pisa, Italy.

PROPERTIES OF STEAM

National Engineering Laboratory, 1964. *Steam Tables*. H.M.S.O.
KEENAN; KEYES; HILL & MOORE. 1969. *Steam Tables*. New York, Wiley. (Metric or English Units.)
Electrical Research Association, 1967. *Steam Tables*. New York, Arnold & St. Martin's Press (Old British Units).
LYLE, O. 1963. *The Efficient Use of Steam*, H.M.S.O.

PRESSURE MEASUREMENTS

American National Standard Institute, 1968. *Gauges Pressure and Vacuum, Indicating Dial Type—Elastic Element*. (B40.1-1968.) New York, American National Standards Institute.
A.S.M.E. Power Test Codes: *Supplements on Instruments and Apparatus*. New York, American Society of Mechanical Engineers.
British Standards Institution, 1780-1960. *Specification for Bourdon Tube Pressure and Vacuum Gauges*. London.
DOOLITTLE, J. S. 1957. *Mechanical Engineering Laboratory Instrumentation and its Application*. Chapters 1, 2. New York, McGraw-Hill.
HUNT, A. M. 1961. The Measurement of Borehole Discharges, Downhole Temperatures and Pressures, and Surface Heat Flows at Wairakei. *Proc. U.N. Conf. on New Sources of Energy, Rome, 1961* (35/G/19). New York, United Nations.
KALLEN, H. P. 1961. *Handbook of Instrumentation and Controls*, Section 3. New York, McGraw-Hill.
WAINWRIGHT, D. K. 1970. Subsurface and Output Measurements on Geothermal Bores in New Zealand. Pisa, 1970.

TEMPERATURE MEASUREMENTS

American National Standards Institute, 1966. *Automatic Null Balancing Electrical Measuring Instruments*. (C 39.4-1966.) New York.

American National Standards Institute, 1964. *Temperature Measurement Thermocouples*. (C 96.1-1964.) New York.
BANWELL, C. J. 1957. Physics of the New Zealand Thermal Area. *Bull. Dep. sci. industr. Res. N. Z.*, no. 123.
BODVARSSON, G.; PALMASON, G. 1961. Exploration of Subsurface Temperatures in Iceland. *Proc. U.N. Conf. on New Sources of Energy, Rome, 1961*, New York, United Nations.
British Standards Institution, 1943. *Temperature Measurement*. (Code 1041.) London, British Standards Institution.
DENCH, N. D. 1962. Reconditioning of Steam Bores at Kawerau. *N. Z. Engng*, vol. 17, no. 10.
Japan Geothermal Energy Association, 1969. *Geothermal Energy in Japan*.
KENT, W. 1950. *Mechanical Engineers' Handbook*. Sections 18, 19. New York, Wiley.
McNITT, J. R. 1963. Exploration and Development of Geothermal Power in California. *Spec. Rep., Calif. Div. Min. Geol.*, no. 75.
MINUCCI, G. 1961. Rotary Drilling for Geothermal Energy. *Proc. U.N. Conf. on New Sources of Energy, Rome, 1961* (35/G/66). New York, United Nations.
SAITO, M. 1961. Known Geothermal Fields in Japan. *Proc. U.N. Conf. on New Sources of Energy, Rome, 1961* (35/G/1). New York, United Nations.
WHITE, D. E. Preliminary Evaluation of Geothermal Areas by Geochemistry, Geology, and Shallow Drilling.
(Also: American National Standard, 1968; British Standards, 1780-1960; DOOLITTLE, 1957; HUNT, 1964; WAINWRIGHT, 1970.)

FLOW MEASUREMENTS

A.S.M.E., 1933-1937. *Fluid Meters* (Theory and Application, 1937; Selection and Installation, 1933). New York, American Society of Mechanical Engineers.
BANGMA, P. 1961. The Development and Performance of a Steam-Water Separator for use on Geothermal Bores. *Proc. U.N. Conf. on New Sources of Energy, Rome, 1961* (35/G/13), New York, United Nations.
British Standards Institution, 1964. *Flow Measurement*. (Code 1042.) London, British Standards Institution.
EINARSSON, S. S. 1961. Proposed 15 MW Geothermal Power Station at Hveragerdi, Iceland. *Proc. U.N. Conf. on New Sources of Energy, Rome, 1961* (35/G/9). New York, United Nations.

International Standards Organization, 1967. *Orifice and Nozzle Measurements*. (R 541-1967 (E).)

JAMES, R. 1962. Steam-water Critical Flow through Pipes. *Proc. Inst. mech. Engrs, Lond.*, 176 (26).

NENCETTI, R. 1961. Methods and Apparatus used for Well-mouth Measurements in the Larderello Zone when a new well comes in. *Proc. U.N. Conf. on New Sources of Energy, Rome, 1961* (35/G/75). New York, United Nations.

PIIP, B. I.; IVANOV, V. V.; AVERIEV, V. V. 1961. The Hypothermal Waters of Pauzhetsk, Kamchatka, as a Source of Geothermal Energy. *Proc. U.N. Conf. on New Sources of Energy, Rome, 1961* (35/G/38), New York, United Nations.

(Also: American National Standard, 1968; British Standards, 1780-1960; HUNT, 1964; DOOLITTLE, 1957; WAINWRIGHT, 1970; BANWELL, 1957; DENCH, 1962; Japan Geothermal Energy Association, 1969; McNITT, 1963.)

MISCELLANEOUS DOWN-HOLE MEASUREMENTS

GATLIN, C. 1960. Petroleum Engineering, Drilling & Well Completions. Prentice-Hall.

Schlumberger Well Surveying Corporation, 1958. *Introduction to well logging*. (Doc. no. 8.) Houston, Texas, Schlumberger.

(Also: WAINWRIGHT, 1970; DENCH, 1962; SAITO, 1964.)

Collection and transmission of geothermal fluids

J. H. Smith

Chief Geothermal Engineer,
Ministry of Works (New Zealand)

1. Introduction

The character of fluids discharged from geothermal wells varies considerably. In some areas only steam (slightly superheated) is produced whereas elsewhere wells yield a mixture of steam and water (i.e. two-phase). In other cases only hot water is produced, either above or below 100 °C: if it is at or above 100 °C sufficient pressure is maintained to suppress boiling and formation of steam. In most cases the fluid produced contains incondensible gases (typically CO_2 and H_2S with minor amounts of other gases), the proportion of gas sometimes being so high that economic utilisation is not feasible. Where water is present it contains a variety of chemicals in solution, the recovery of which may be profitable. When first discharged after drilling, rock fragments are usually ejected; sometimes in great quantity. Eventually wells blow themselves clear of such debris but small particles may continue to be entrained in the fluid produced. All of these impurities present some problems in utilisation. Water contained in steam can also be regarded as an impurity but in many cases the heat in the water can be used to good effect. The water/steam ratio may be quite high, and ratios of 4/1 to 8/1 (by weight) are not uncommon. Varying the production pressure of a particular well affects the water/steam ratio, a lowering of pressure causing an increase in total flow and an increase in the proportion of steam.

Before designing a collection and transmission system (and indeed the plant required for utilisation) it is necessary to know the chemical properties of the fluids produced from the wells, the effect of varying the wellhead pressure on the rate of flow and the water/steam ratio, and the maximum wellhead pressure in closed wells. Optimisation studies will also have been made to determine the economic pressure at the point of utilisation and the operating pressure at wellheads.

In a typical geothermal field there will be a number of wells connected to one or more pipelines leading to the plant which may be located a considerable distance away. The fluid transmitted may be steam, hot water, or a steam/water mixture if separation is effected at the plant instead of at the wellheads. Furthermore some fields can be best exploited by operating groups of wells at different pressures with a separate transmission system for each group.

2. Wellhead equipment

The wellhead pressure in a closed well may be very high and much in excess of that occurring under operating conditions when the flow is controlled. This is particularly so in wells which produce steam/water mixtures with which gas is associated. When standing shut, the gas may accumulate at the top of the well and depress the water column until pressure equilibrium is established. In such instances wellhead pressures of several hundred lb/in^2 have been recorded. This pressure can be relieved by a continual small discharge to waste through a bleed valve installed below the wellhead master valve.

In addition a shut-off valve may have an abnormally high force imposed on the gate during closure due to impact of the steam/water mixture moving at high velocity. To prevent damage, valves used for this purpose should have large seating surfaces and be well guided to prevent shearing of mating surfaces. It is good practice to operate the master valve only under conditions of no flow and to provide other valves downstream for flow control. Should malfunction of the latter occur the master valve is then available as a last resort.

Wells producing steam only can be connected direct to the steam transmission system, but where steam and water require to be separated at the wellheads various arrangements are possible depending on whether the separated water is to be wasted or piped away for further use.

If the water is to be wasted, an arrangement of equipment such as that shown in Figure 1 is suitable. Separation takes place in a cyclone separator from which the steam flows direct to the steam mains, the pressure in which controls the separation pressure. Water from the separator

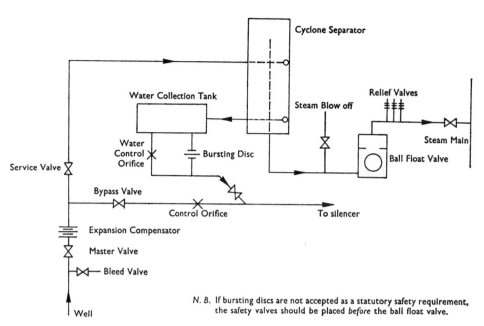

N. B. If bursting discs are not accepted as a statutory safety requirement,
the safety valves should be placed *before* the ball float valve.

FIG. 1. Wellhead arrangement.

goes to a collection tank in which the water level is control-led by a water control orifice to prevent either carry-over of steam by loss of seal or flooding of the separator and entry of water into the steam main. Suitably shaped water control orifices when discharging boiling water have self regulating features and can pass a wide range of flows (about 2½/1 ratio) without loss of seal or flooding. The separated water is discharged to waste through a silencer; but if desired, steam at a lower pressure could be obtained from it by flashing and passing it to a second stage separator before final discharge to waste.

The provision of a ball float valve in which the ball would rise and shut off the flow is a safeguard against possible carry-over of water into the steam main. The resulting increase in pressure would cause a bursting disc to rupture and release the flow to the silencer. Under normal operating conditions safety valves provide protection in the event of undue pressure rise in the steam mains.

A bypass line provides for the total flow from the well to be discharged to waste, the flow and pressure being controlled by an orifice of suitable size inserted in the pipe. (This is preferable to throttling with the bypass valve.) Bypassing of the flow enables the wellhead equipment to be isolated for inspection and maintenance without closing the well. It is also a means by which the flow can be stabilised before diverting it through the separator by manipulating the valves. During this process and before allowing it to enter the steam main, steam from the separator is discharged through the blow-off pipe until complete separation is accomplished. The bypass line is, of course, installed soon after a well is drilled to enable output tests to be made and the behaviour of the well over a lengthy period to be observed.

Separation of steam and water is effected by centrifugal action in a vertical cylindrical vessel into which the mixture enters tangentially through a spiral inlet pipe (see Fig. 6). The water is forced to swirl around the wall and gravitates downwards, leaving the vessel through a tangential outlet near the bottom. The steam first flows upwards and leaves through a central pipe emerging from the bottom. In an alternative design steam is taken off from the top of the vessel. In a well designed separator the efficiency of separation is high, residual water in the steam being less than one part per thousand by weight.

If waste fluids are allowed to escape to the atmosphere through an open ended pipe the noise created can be a nuisance and also damaging to the hearing of people working in the vicinity for extended periods. It is therefore necessary to abate the noise by some form of silencing and at the same time to control the flow of waste water.

The most effective method is to discharge the effluent underwater but it is of limited application unless close to a lake or river. In suitable locations an artificial pond could be formed if the quantity of waste is small, surface cooling of the pond and evaporation being relied on to dissipate the heat. For temporary or intermittent use, horizontal silencers in which the pipe diameter is gradually increased are useful but a disadvantage is difficulty of controlling the discharged water. One type consists of a steel cone welded to a short cylindrical section at the discharge end. A similar effect can be obtained by use of concrete pipes of increasing diameter laid horizontally and partly telescoped one inside another.

For a permanent installation a vertical twin cyclone silencer, as used in New Zealand and elsewhere, is very efficient (see Fig. 2). The entering stream of steam and water

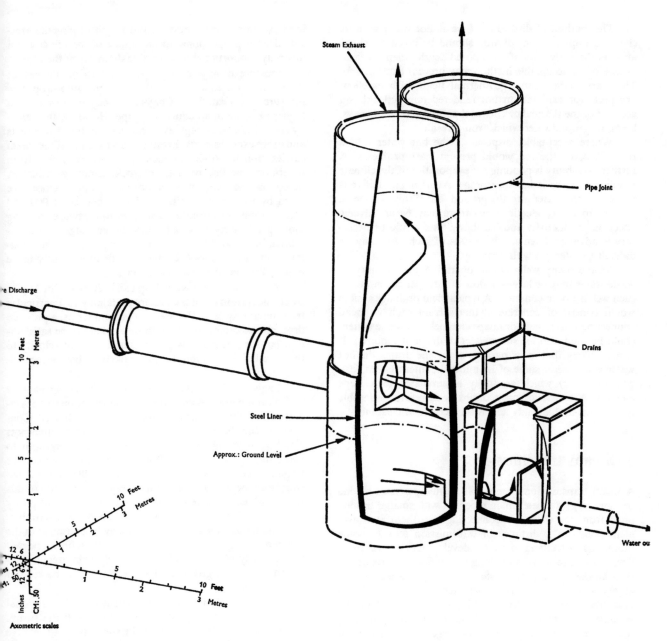

FIG. 2. Twin Cyclone Silencer.

is bifurcated on a steel flow splitter and the separate streams flow in clockwise and anticlockwise rotational swirls. The steam expands rapidly and emerges at low velocity from the stacks while the swirls of water impinge and discharge to a chamber with a weir which controls the outflow and the water level in the bottom of the silencer. Most of the kinetic energy of the entering fluid is destroyed by friction and turbulence. For the construction of the stacks wooden stave pipes have proved to be more satisfactory than reinforced concrete or concrete lined steel pipes as very little

maintenance is required. The staves are made from radiata pine timber preservatively treated with pentachlorphenal.

Sound levels as high as 140 decibels have been measured for a discharge from the end of a straight pipe whereas the intensity can be reduced to 100 decibels by provision of an efficient silencer. (A reduction of 3 decibels represents an approximate halving of the intensity.) In addition to reducing the noise level much of the noise with a frequency in the region of 1,000 cycles per second, which is particularly unpleasant, is eliminated.

The method of disposing of waste hot water with its chemical impurities in solution should be given consideration in the early days of a proposed development project, as lack of an acceptable method may prevent exploitation. There are in fact some geothermal fields with excellent prospects for exploitation but rendered impracticable on account of the thermal or chemical pollution of waterways, lakes, or groundwater which would result.

Where acceptable, disposal of the hot water into a river, lake, or the sea would present least problems. A further possibility is ponding for evaporation if the climate is suitable or for indefinite storage. Another alternative is to reinject the water into the ground, but if this is done in or near to the production area there may be an adverse effect on productivity and feasibility tests would be necessary in advance. Even so, the results of such tests may be difficult to interpret with certainty.

Where many wells are in operation the quantity of waste water may be large—a flow of 2 ft^3 per second from each well is not uncommon. A typical field drainage system would consist of concrete channels from each wellhead connecting into a main drainage channel. The same system could be used also for stormwater drainage if required. It is most likely that in its passage along the drains the hot water will deposit some of its mineral content (silica, calcite, or other) which gradually accumulates to a large amount. A system of open drains is therefore preferable to pipes to enable ready access for periodical cleaning.

3. Steam transmission

A steam transmission system will normally consist of one or more large diameter main pipelines with smaller branch pipes from the wellheads connecting into them at convenient points. The distance of transmission may be some thousands of feet and in a large development several pipelines may be required stretching into different parts of the field. In the design of the system one of the main considerations is the choice of pipe diameter so that the drop in pressure between wellheads and the delivery end of the pipeline is not excessive. The choice of operating pressure will naturally depend on the discharge characteristics of the group of wells and their productivity at various wellhead pressures. Pressure drop, which is mainly due to frictional resistance to flow, causes the energy in the steam to be degraded and, if excessive, imposes an undesirable pressure increase at the wellhead and consequent reduction of output.

Also contributing to pressure drop are the losses due to disturbance of steady flow created by valves, bends and other fittings, and by sudden enlargements or contractions of diameter. These losses can be assessed in accordance with normal practice.

Having decided on an acceptable pressure drop and flow rate the diameter of a pipeline can be calculated using well known properties of steam and friction factors affecting flow. For a given pressure and length of pipe the pressure drop is proportional to the square of the velocity and inversely proportional to the inside diameter of the pipe.

Pressure drop is also proportional to the density of the steam. Consequently, for the same pressure drop, low pressure steam can flow at higher velocity than steam at a higher pressure in an equivalent pipe. However, the velocity should not be so high as to cause erosion of valve seats and other exposed parts. Erosion is accentuated if the steam carries moisture (due to condensation or carry-over from separators) or fine particles of rock. Limiting velocities, based on the experience of normal steam practice, are given by various authorities and vary between 6,000 and 12,000 ft/min for saturated steam, the lower figure applying more particularly to small pipes. Some judgement is required, but 12,000 ft/min should be regarded as the maximum for larger pipes (12″ diameter or more) likely to be required for geothermal steam service.

The allowable pressure drop can only be decided upon by consideration of all the factors pertaining to a particular transmission system. However, in general terms, a pressure drop not exceeding about 30 lb/in^2 is likely to be satisfactory, but for steam at low pressure limitation of velocity to the permissible maximum is more likely to be the controlling factor.

Lagging of pipes and vessels with an insulating material will minimise heat losses, but some condensation will normally occur unless the steam is sufficiently superheated. Condensate can be removed by collecting it in catch-pots installed in the pipeline at intervals and discharging it to waste through steam traps. With separated steam it cannot be guaranteed that some salt-laden water will not be carried over from the wellhead separators; repeated dilution of this water with condensate and partial removal serves to purify it in transit. The pipeline thus acts as an efficient scrubber. Any small quantity of water which may be entrained in the steam reaching a turbine would then contain only a negligible quantity of sodium chloride which, if in greater concentration, could give rise to stress corrosion cracking of turbine blades.

Due to malfunction of equipment or an operational error, water could enter the mains along with the steam, and in excess of the capacity of the steam traps to remove it. As a slug of water reaching a turbine could have disastrous effects, the installation of water detectors in the mains is a safety precaution. These could be float switches installed in the steam trap catch-pots to detect excess water level and to transmit a signal to warn the operator and initiate sequential tripping of the turbo alternators.

As noted above, ball float valves at the wellheads should normally prevent carryover of water from separators in the event of their flooding.

Under operating conditions it is normal for wellhead valves to be fully open and pressure regulation to be controlled at the plant. The pressure at each wellhead will thus be higher than the controlled pressure by an amount equal to the pressure drop between the wellhead and the plant, and the flow from each well will adjust itself in accordance with its own pressure/flow characteristics. In an extreme case the minimum operating pressure of a well may be abnormally high and necessitate throttling at the wellhead.

So that sufficient steam will be available to meet the maximum demand, somewhat more than the minimum number of wells should be connected to the system, as it is not possible for a certain number of wells to give the exact quantity required. The excess steam would be discharged to waste through a vent valve which would also serve to accommodate fluctuations in the demand for steam. A reduction in the demand would tend to cause a rise in pressure and lifting of safety valves at the wellheads with consequent erosion of their seats. Also, safety-valves may be designed not to reseat until the pressure has fallen below normal. Use of the vent valve to discharge all excess steam overcomes this disability, the degree of valve opening being controlled by a pressure-sensitive device operating over a small range of pressure.

4. Hot water transmission

When hot water at or near boiling point is to be transmitted through a pipeline, precautions are necessary to ensure that the pressure at all points in the pipe is higher than that corresponding to the boiling point so that formation of steam is suppressed. If the pressure were lower, steam would form as bubbles or pockets in the water, and if a pressure rise should subsequently occur collapse of the steam could cause serious water hammer and possible rupture of the pipe, with disastrous consequences. Hot water at moderately high pressure and temperature should be regarded as explosive—much more so than an equal volume of steam at the same temperature and pressure in an equivalent pipe or vessel: at 200 ºC, for instance, it is about 12 times more explosive. To maintain an adequate margin of pressure to suppress boiling, pumping can be employed to increase the pressure, or the temperature can be decreased to below the boiling point by introducing cooler water (attemperation). A combination of both methods may be possible.

Due to friction there is a gradual decline in pressure during flow along a pipeline and a further decline due to loss of hydrostatic pressure if it is on a rising gradient. If the gradient is falling, the hydrostatic pressure increases and may more or less counterbalance the frictional pressure loss, but in practice the gain in pressure may be insufficient to compensate losses when surge effects are also taken into account.

Surging occurs when the velocity changes, as may result from operating a control valve at the delivery end of the pipeline; and pressure changes then occur at all points along the pipeline. Every movement of the valve, opening or closing, causes pressure oscillations, the magnitude of which depends on the rate of velocity change. It can readily be imagined that in a long pipeline a rapid retardation of the mass of moving water would cause a large rise in pressure; but more important is the pressure decline (negative surge pressure) caused by rapid acceleration with resultant formation of steam which subsequently collapses. Changes in velocity caused by rate of valve movement or otherwise must therefore be carefully controlled to restrict resulting surge pressures within acceptable limits.

Pressurisation by pumping involves installing a pump at each wellhead and pumping the water separated at boiling point into the pipeline. To stabilise the pumping pressure and to provide storage capacity to take up transient changes of flow along the pipeline or from the wells it is convenient to provide a head tank connected to the pipeline, the water level in this tank being controlled, within limits, at an elevation above that of the highest well. By connecting the vapour space above the water in the tank to the steam line (operating at the same pressure as that at which the water is collected by the pumps) the hot water pipeline is subjected to an excess pressure dependent on the elevation of the head tank.

An example of hot water transmission is provided by a pilot plant which was installed at Wairakei in New Zealand. The scheme was designed for five high pressure wells and two intermediate pressure wells operating at approximately 200 lb/in² and 70 lb/in² respectively. The length of the pipeline was about 5,000 ft and it delivered the hot water to a flash plant located alongside the power station, flash steam being produced in two stages at pressures of 50 lb/in² and just above atmospheric. The scheme is illustrated in Figure 3 which shows only one of each class of well connected to the hot water main.

In this case attemperation in conjunction with pumping was adopted, the IP[1] water being pumped into the HP[2] water system. The reduction in temperature thus effected enabled a significant reduction to be made in the maximum pressure to which the water was pumped and in addition some of the heat in the IP water became available for use. At each wellhead, water passed from the separator to a collection vessel from which it was pumped to the hot water main, a relief valve on the collection vessel allowing water to spill to waste should the supply of water exceed the capacity of the pump. At the delivery end of the pipeline the hot water passed through automatic control valves which let down the pressure to that of the IP flash vessels. The rate of movement of these valves was restricted to ensure that surge heads were kept within safe limits.

Vapour pumping is another method of pressurising which has been suggested. It is shown diagramatically in Figure 4 for application where two separate steam pressure systems are available but it could be operated less efficiently by using the atmosphere as the lower pressure system. As for the scheme described above a similar hot water collec-

---

1. Intermediate pressure.
2. High pressure.

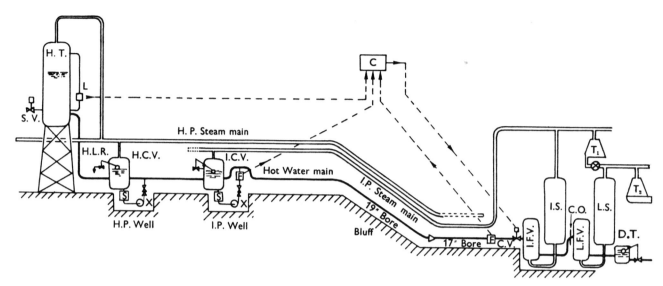

FIG. 3. Hot water transmission system.

Key:

H.C.V.	H.P. collection vessel	C.V.	Motor operated control valve
I.C.V.	I.P. collection vessel		
H.T.	Head tank	I.F.V.	I.P. flash vessel
S.V.	Motor operated spill valve	L.F.V.	Low pressure flash vessel
L.	Level detector	I.S.	I.P. Scrubber
F.	Flowmeter	L.S.	L.P. Scrubber
H.L.R.	High level relief	D.T.	Drain tank
S.	Strainer	T_1	I.P. turbine
X.	Extraction pump	T_2	L.P. turbine
C.	Computer	C.O.	Control orifice
==	Steam connections	H.P.	High pressure
——	Hot water connections	I.P.	Intermediate pressure
- - -	Electrical connections	L.P.	Low pressure

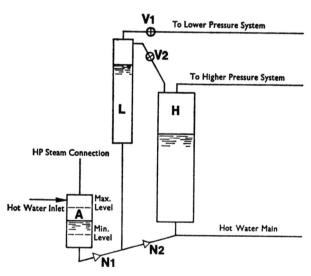

FIG. 4. Vapour pumping.

tion vessel 'A' is provided for each well and a common head tank 'H' for the whole group of wells. In addition a lift cylinder 'L' is associated with each collection vessel. Valves V1 and V2 are controlled by signals when the water in vessel 'A' reaches its maximum and minimum levels. N1 and N2 are non-return valves.

The sequence of operation is as follows:

1. V1 closes and V2 opens.
2. Water in 'A' rises to maximum level.
3. V1 opens and V2 closes.
4. Excess pressure in 'A' forces water into 'L' until minimum level in 'A' is reached.
5. V1 closes and V2 opens, equalising pressures above the water in 'L' and 'H'.
6. Cycle recommences and water drains from 'L' until levels in 'L' and 'H' equalise.

While less efficient than mechanical pumping vapour pumping would have the advantage of extreme mechanical simplicity and several wells at different ground levels could be connected into the same system. With mechanical pumping every wellhead pump would have different characteristics to cope with differences of ground level, different flows from wells and variation of these flows.

5. Two phase transmission

For a scheme in which it is intended to utilise the hot water, some advantages may be gained by transmitting in one or more pipelines the mixture of steam and water as produced from the wells with separation taking place at the point of utilisation. Much of the equipment at wellheads would then be eliminated including separators, hot water pumps, water collection tanks, and all the associated pipework. As there would be no need to suppress boiling in the pipeline a head tank on the hot water line would also not be required. Even if the hot water were not to be used the scheme may still be worthwhile since permanent silencers at the wellheads would not be required and the water would be rejected at one point after separation at the delivery end of the pipeline, thus avoiding the need for an extensive system of surface drains.

After primary separation the hot water would be rejected or would undergo a reduction in pressure in a flash plant to produce steam in one or more stages at lower pressures with final rejection of residual hot water.

The character of the flow in a pipeline of a mixture of steam and water assumes different forms depending on the volume of each fluid present and on the velocity. As with any mixture of a liquid and gas (or vapour) it is quite different from the flow characteristics of each of the phases if it alone occupied the pipe. It can best be understood by considering a pipe containing hot water under pressure into which steam at the same temperature and pressure is introduced in progressively greater amounts. The type of resulting flow would be generally as follows:

Bubble Flow: Bubbles of steam move along the upper part of the pipe at approximately the liquid velocity.

Plug Flow: Alternate plugs of steam and water move along the upper part of the pipe above the water.

Stratified Flow: Water flows in the bottom of the pipe with the steam above over a smooth interface.

Wave Flow: Due to higher velocity the interface has become disturbed by waves travelling in the direction of flow.

Slug Flow: A slug of water is picked up periodically by the steam moving more rapidly and continues at much greater velocity than the mean velocity of the water.

Annular Flow: The water forms in a film around the inside of the pipe and the steam flows at high velocity in a central core.

Spray Flow: Nearly all of the water is entrained in the steam.

The design and operation of a two phase steam/water transmission system presents a number of problems. Most of the experimental and operational information available is limited and confined to comparatively short pipelines.

The calculation of pressure drop, which is vital for determining the pipe diameter, is much less precise than for either steam or water flowing in separate pipes although an approximation can be made by use of data developed primarily for oil/gas mixtures. The pressure drop due to friction will be greater for a steam/water mixture than for the same quantity of steam only in an equivalent pipe. Added to this are the rather indefinite pressure drops which will occur at bends, tees, manifolds, etc.

There is also the possibility of slugs of water forming in the pipeline and being carried along at high velocity to cause impact at bends, etc. This can be minimised by ensuring that the nature of the flow is in the annular or spray flow region. In practice this is likely to be the case as the mixture velocity will be high enough to ensure this. Even so, under some conditions of operation the velocity may be much less and cause an undesirable flow regime to develop.

With the type of two-phase flow under consideration the steam/water ratio is such that it would be possible for only a fraction of the steam to be condensed by increase of pressure. Complete collapse of the steam would not therefore occur, as could be the case with transmission of boiling water only, and problems associated with severe water hammer would be virtually eliminated.

6. Pipework practice

Geothermal fluids are potentially corrosive but tests and practical experience have shown that in the absence of oxygen mild steel has good corrosion resistance and is suitable for wellhead vessels, steam and hot water pipes, and the like. Other materials such as stainless steel may be necessary where abnormal corrosivity is present.

Steel pipe is made by one of several processes which produce either seamless pipes or welded pipe. Seamless pipe is manufactured by hot working steel to produce the required dimensions and properties, followed if necessary by cold finishing of the hot worked product. Welded pipe has a longitudinal seam formed by fusion of abutting edges after rolling plate or strip to cylindrical form. Fusion may be obtained by heating in a furnace or by electrical means and by application of mechanical pressure to force the edges together. The more commonly understood type of welded pipe is that in which the buttjoint is made by electric arc welding, either manual or automatic, with the use of filler material. It is particularly suitable for large diameter pipe and if the welding is properly controlled the strength of the weld approaches that of the plate itself. Another type of pipe is spiral welded in which steel strip is rolled into the form of a spiral and the butting edges welded together.

Pipe is normally purchased to comply with a standard specification applicable to the purchaser's conditions of use, some options being specified by the purchaser. Such a specification includes the quality of the steel and its minimum strength properties, the method of manufacturing the pipe, its physical dimensions, and the tests necessary to ensure compliance with the standard. If electric arc welding is used only qualified welders and procedures should be employed and for high quality work, in addition to physical tests on test specimens of production welds, each weld should be examined by radiography and defects repaired.

Failure to provide for axial expansion of installed pipes due to temperature change may set up very high stresses with consequent damage to the piping, anchors and supports. With long runs of pipe, expansion may be several feet between hot and cold conditions—for instance, about 2 ft expansion per 1,000 ft of pipe carrying steam at a pressure of 150 lb/in².

Standard piping practice provides for axial expansion to take place between anchors to which the pipe is fixed and which transfer thrusts to the ground or to a solid structure. Main anchors are normally provided at pipe ends, at changes in direction, at shut off valves, at manifolds where pipes are interconnected, and elsewhere as dictated by the expansion arrangements adopted. Intermediate anchors may also be provided to divide the pipeline into separate expanding sections and to bear any unbalanced thrust. Pipe movement is accommodated by supporting the pipe on rollers or slides or by use of flexible

hangers from supporting structures. To prevent buckling either horizontally or vertically guides may also be necessary. Branchline connections should preferably be made at or near anchor points on main pipelines.

In many instances advantage can be taken of the inherent flexibility of the pipe to limit expansion forces to acceptable figures. Situations which lend themselves are properly proportioned runs of pipe intersecting at an angle or where directional changes can be deliberately introduced such as a zig-zag arrangement. Where this is impracticable expansion bends of U shape, rectangular loops, or other configurations may be inserted into straight runs of pipe. Relief of thermal stress by flexing is accompanied by smaller stresses due to bending in the pipe and bends and free movement to allow such flexing must be provided. With large diameter pipe the radii of U bends and the offset length of rectangular loops may be inconveniently large.

Other methods of providing for expansion include the installation of slip joints or bellows expansion compensators. With a slip joint, purpose made as a unit for insertion into a straight run of pipe, a sleeve slides inside a larger diameter body and relies on a packing and stuffing box for fluid tightness. Within limits, any desired expansion can be accommodated, but due to leaks occasional maintenance may be required. Furthermore they would not be suitable with geothermal fluids likely to deposit minerals.

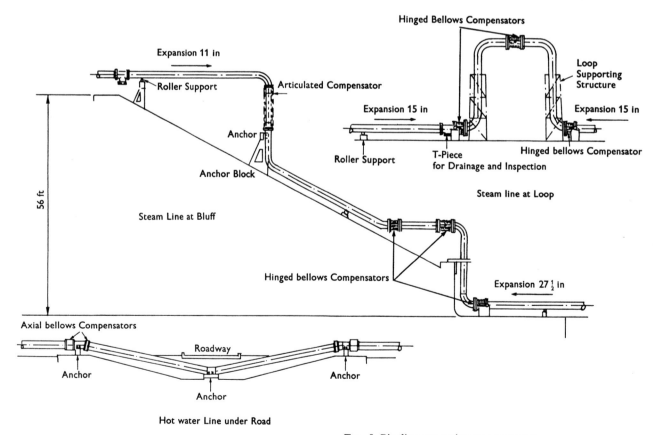

FIG. 5. Pipeline expansion arrangements.

Bellows expansion compensators are available in a variety of types—axial, angular (hinged), articulated and gimbal. Basically they consist of a corrugated metal cylinder (typically stainless steel). With the axial type expansion is accommodated by compression of the convolutions, internal liners being provided where high velocities are encountered or to protect the convolutions from particles of sediment which may become trapped in them and prevent them from closing. With the other types the convolutions are compressed on one side and elongated on the opposite side, the effect being to introduce a hinge into the pipeline. Devices to prevent excessive stressing of the convolutions are incorporated when necessary to limit the amount of compression or angular movement. By a suitable combination and placement of compensators it is possible to provide for expansion in practically any configuration of pipework. Figure 5 shows some examples. In a long straight pipe the introduction midway between anchors of loops with hinged compensators placed as shown, enables a large amount of expansion to be accommodated—for maximum permissible angular movement of the compensators the amount of expansion is governed by the length of the parallel legs of the loop. Vertical loops provide conveniently for a roadway beneath.

In the fabrication and erection of pipework joints may be made by bolted flanges or by circumferential welding. Flanges, which may be attached to pipe ends either by screwing or by welding, are required where disassembly is necessary for maintenance work or inspection and where flanged components are incorporated in the system. Elsewhere joints made in the field by welding together adjacent pipe ends are entirely satisfactory if the work is properly controlled. Pipe ends should be prepared for butt welding, only qualified welders should be employed, and welding equipment and procedures should comply with recognised practice. While not always mandatory it is most desirable that each weld be examined by radiography and any defects repaired. Heat treatment to relieve stress in the welded joint may be required if the pipe is more than $\frac{1}{2}''$-$\frac{3}{4}''$ thick.

In designing a piping system it is usual to provide for pipe to be stretched between anchors during erection by an amount equal to approximately half of the calculated thermal expansion. When the line comes into service and is gradually heated the tensile stress initially induced is progressively relieved until the zero point is reached and further

heating then induces compressive stress. By this means the ultimate thermal stress is reduced to approximately one half of what it would be if this cold pull-up was not applied. The result is that the load transmitted to anchors is similarly reduced, or if in fact expansion is allowed to take place the amount of expansion is approximately halved.

In a previous paragraph reference was made to the provision of steam traps in a steam line to remove condensate. In addition to regular spacing of these in a long straight run of pipe they should also be installed where condensate is likely to collect at low points in the line.

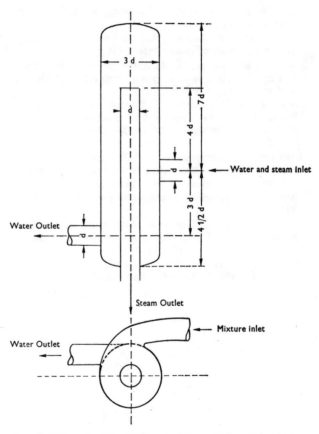

Fig. 6. Diagrammatic sketch of bottom outlet cyclone separator, with approximate proportions.

Bibliography

(a) Papers presented to the United Nations Conference on New Sources of Energy held in Rome, August 1961, and published by United Nations, New York, 1961, as Proceedings of that Conference—Volumes 2 and 3, Geothermal Energy.

ARMSTEAD, H. C. H. 1961. Geothermal power development at Wairakei, New Zealand. Paper G4, vol. 3.

BANGMA, P. 1964. The development and performance of a steam/water separator for use on geothermal bores. Paper G13, vol. 3.

DENCH, N. D. 1961. Silencers for geothermal bore discharge. Paper G18, vol. 3.

DI MARIO, Pietro. 1961. Remarks on the operation of the geothermal power stations at Larderello and on the transportation of geothermal fluid. Paper G68, vol. 3.

EINARSSON, Svein S. 1961. Proposed 15 Megawatt geothermal power station at Hveragerdi, Iceland. Paper G9, vol. 3.

ENGLISH, Earl F. 1961. Methods and equipment for harnessing geothermal energy at The Geysers, California. Paper G51, vol. 3.

SMITH, J. H. 1961. Harnessing of geothermal energy and geothermal electricity production. General report. GR/4 (G), vol. 3.

(b) Papers presented to the United Nations Symposium on the Development and Utilisation of Geothermal Resources held in Pisa, September 22-October 1, 1970. To be published by the Istituto Internazionale per le Ricerche Geotermiche, Lungarno Pacinotti 55, Pisa, Italy.

POLLASTRI, G. 1970. Design and construction of steam pipelines. Paper VIII/1.

ARMSTEAD H. C. H.; SHAW, J. R. 1970. The control and safety of geothermal installations. Paper VIII/3.

JAMES, R.; McDOWELL, G. D.; ALLEN, M. D. 1970. Flow of steam-water mixtures through a 12-inch diameter pipeline. Test results. Paper VIII/4.

TAKAHASHI, Y.; HAYASHIDA, T.; SOEZIMA, S.; ARAMAKI, S.; SODA, M. 1970. An experiment on pipeline transportation of steam-water mixture at Otake Geothermal Field. Paper VIII/5.

UCHIYAMA, M.; MATSUURA, S. 1970. Measurement and transmission of steam in Matsukawa Power Plant. Paper VIII/6.

YANAGASE, T.; SUGINOHARA, Y.; YANAGASE, K. 1970. The properties of scales and methods to prevent them. Paper VIII/7.

OZAWA, T.; FUGJII, Y. 1970. A phenomenon of scaling in production wells. The geothermal power plant in the Matsukawa Area. Paper VIII/8.

NISHIWAKI, N.; HIRATA, M.; IWAMIZU, T.; OHNAKA, I.; OBATA, T. 1970. Studies of Noise reduction problems for electric power plant utilising geothermal resources. Paper VIII/9.

WIGLEY, D. M. 1970. Recovery of flash steam from hot bore water. Paper VIII/11.

JAMES, R. 1970. Collection and Transmission of geothermal fluids—Section VIII. Report by Rapporteur.

(c) Other references.

American Society of Mechanical Engineers, 1967. *U.S.A. Standard Code for Pressure Piping.* (USAS B31.1.0) New York, American Society of Mechanical Engineers.

ARMSTEAD, H. C. H. 1968. The extraction of power from hot water. *Seventh Wrld Pwr Conf., Moscow.*

BAKER, O. 1954. Design of pipelines for the simultaneous flow of oil and gas. *Oil Gas J.,* 53, 185.

BENJAMIN, M. W.; MILLER, J. G. 1941. The flow of saturated water through throttling orifices. *Trans. Amer. Soc. mech. Engrs,* 63.

BOTTOMLEY, W. T. 1936-37. Flow of boiling water through orifices and pipes. *Trans. N. E. Cst Instn. Engrs. Shipb.,* 53.

BURNELL, J. G. 1947. Flow of boiling water through nozzles, orifices and pipes. *Engineering, Lond.,* 164.

HALDANE, T. G. N.; ARMSTEAD, H. C. H. 1962. The geothermal power development at Wairakei, New Zealand. *Proc. Instn. mech. Engrs, Lond.,* 176.

JAMES, R. 1968. Pipeline transmission of steam-water mixtures for geothermal power. *N.Z. Engng,* 23 (10) 55.

LOCKHART, R. W.; MARTINELLI, R. C. 1949. Proposed correlation of data for isothermal two-phase, two-component flow in pipes. *Chem. Engng Progr.,* 45, 1.

SMITH, J. H. 1958. Production and Utilisation of geothermal steam. *N.Z. Engng* 13 (10) 354.

——. 1962. Power from geothermal steam at Wairakei. *Sixth Wrld Pwr Conf., Melbourne.*

TRETHOWEN, H. A. 1960. The flow of gas-liquid mixtures in pipes. *N.Z. Engng* 15 (5) 151.

IV

The utilisation of geothermal fluids

Geothermal power

Basil Wood
Consulting Engineer
(United Kingdom)

This section deals with power generation. Although geothermal energy can be, and is, used for district heating, industry and other purposes, interest has largely been concentrated upon power generation. The reason is that geothermal resources have to be exploited where they occur. These locations are commonly remote from large centres of population. Generation of electricity makes possible the transport of the power over overhead lines to the towns where it is required.

1. Power plants installed

Table 1 contains a list of major geothermal plants in operation or in construction. In a few places, e.g. Salton Sea and others (not included) geothermal plant has been installed but has fallen into disuse.

There are two types of geothermal field, namely those yielding dry steam (Larderello and the Geysers) and those yielding wet steam—a mixture of boiling water and flash steam as at Wairakei, Mexicali, Iceland, and El Salvador. The mechanism of the former type is not fully understood. The latter mechanism is obvious, namely the existence of a reservoir of hot water at or near boiling point under the pressure of the overlying rock. When the pressure is released by issue through a drill hole a part of the water flashes into steam. This process can readily be calculated by elementary heat balance. As a rough rule 1 % of the water will flash into steam for every 10 °F reduction in saturation temperature. In practice the proportion of steam to water obtained in various fields is 1 : 4 to 1 : 10, the steam being separated from the water in cyclone separators ordinarily at the wellhead. The boiling water is ordinarily discarded. In either case the steam plant differs from fuel fired plant in that it has to utilize nominally saturated steam without significant superheat. The 'dry' steam at Larderello and the Geysers may by throttling acquire a few degrees of superheat but the expansion through the turbine is effectively entirely in the wet region. At the beginning such turbines were unique in this respect but now can be regarded as similar to those in water moderated nuclear power stations so some of the problems in using wet steam have become familiar. However there are differences as mentioned in the next paragraph and in Section 9 later.

TABLE 1. Geothermal power stations

Plant	Country	Date	Total installed MW	Largest unit MW
Larderello	Italy	1930-1969	330	26
Wairakei	New Zealand	1958-1963	198	30
Kawerau	New Zealand	1961	10	6
The Geysers	California	1960-1971	192[1]	55
Salton Sea	California	1966	3	1.5
Pauzhetka	Kamchatka	1967	5	2.5
Paratunka	U.S.S.R.	1970	0.5	0.5
			(Freon)	
Matsukawa	Japan	1966	20	20
Otake	Japan	1967	11	11
Pathé	Mexico	1958	3.5	3.5
Mexicali	Mexico	1972?	75	37.5
Akureyri	Iceland	1969	2.5	2.5

1. Plus 220 more visualized.

The wide application of wet steam in American nuclear plants has greatly stimulated interest in the subject and influenced both attitudes and equipment available. For instance it has created much greater interest in separators. The influence is not entirely beneficial since in nuclear engineering, because of heavy subsidies and deliberate under-estimating of costs, very expensive techniques have been employed such as resort to stainless steel on every occasion or at least stainless steel cladding. Geothermal engineers, being confined to much more moderate budgets, have been obliged to find cheaper solutions.

2. Contaminants in steam

Geothermal saturated steam differs from that produced in nuclear plants in containing various contaminants. First all geothermal steam seems to contain the gases CO_2, H_2S, and NH_3 (and others of lesser consequence). These may appear in different proportions and to a total of between $\frac{1}{3}\%$ and 10% or more. There is ordinarily no oxygen and in its absence H_2S is not corrosive to steel (or at least not after the steel assumes a protective patina of FeS).

The water phase also normally contains common salt (NaCl) and silica, and may contain scale forming salts. Common salt may be about 3,500 ppm (about 1/10 the strength of sea water) or higher. Silica is likely to be in saturation at the depth of origin—therefore in supersaturation at atmospheric pressure. Silica is also sparingly soluble in the steam phase, more so at higher pressure. Its deposition consequent on a drop of temperature is possibly beneficial in some cases as providing a protective skin. However it can also hinder the movement of valve rods and hence presents a menace to the safety of the plant.

Since chlorides are particularly undesirable from the corrosion aspect, particularly in the turbine blades, it is important to get rid of the traces of water carried over from the initial separation. This is done conveniently in a long steam pipe since condensing steam dilutes the initial salty water and the condensate is discharged through drains. Dry steam as at the Geysers may also contain fine pumice dust. This is not ordinarily a problem with wet steam unless the separation is very inefficient, but wet wells if allowed to blow violently or not provided with slotted casing may deliver substantial quantities of coarse rock particles.

3. Basis of selection of materials for geothermal schemes

Initially common constructional materials were adopted which were already in use in other plant and experience was obtained through failures. Practically the whole of the materials required at the wells—drill rods, casings, etc.—had already been evolved in long experience of oil well drilling where sour wells, i.e. those containing sulphur, were already known to produce catastrophic failure in high strength alloys. The general principle that the more superior the alloy in regard to high strength properties the worse it behaved, was already recognised. A good deal could be deduced from careful inspection of parts such as casing and valves after exposure for some time to geothermal fluids. Early experience showed severe cutting of valve seats and faces wherever leakage of wet steam took place. Cutting by rock particles also occurred which was later avoided by suitable devices. Many valves accepted in other industries were found to be incapable of making a tight seal or closing against the flow under pressure. Stelliting of faces was the remedy and use of stainless steel trim.

By the time large geothermal turbines started to be built, evolution of blade materials had already settled on 12% Cr 'iron', i.e. low carbon, as the best all-round material and this (even with higher carbon) was found to stand up to H_2S so long as it was not in hardened (martensitic) state, e.g. through either deliberate or inadvertent heat treatment such as was liable to occur by application of a torch flame for instance in fixing lacing wires by brazing.

Before it became known that geothermal steam has generally similar properties in all of the fields where it is exploited, it was thought essential to carry out tests on coupons of a wide range of materials both for general corrosion resistance and for stress corrosion cracking. Appropriate procedures had already been developed in the oil business. Such tests required however interpretation and it was possible to derive misleading ideas. For instance the chemist removes the corrosion film and deduces the loss in weight. He thus exaggerates the corrosion rate in that the initial attack may well produce a protective film which would have greatly slowed down the subsequent rate of attack. This is indeed the case with ferrous materials exposed to H_2S at least in the absence of oxygen. Similarly test coupons are commonly chemically cleaned before exposure. This tends to exaggerate corrosion rate since quite small films of oil will confer some protection.

Short period tests are all that ordinarily can be undertaken usefully in the few months available between the decision to go ahead with the project and the drawing up of the specification. One cannot wait 20 years which is the nominal life of the plant for tests either on general corrosion resistance, erosion, or on stress corrosion cracking and corrosion fatigue. Consequently accelerated techniques must be adopted in order that data can be obtained in a period of less than a year. This means for instance overstressing beyond the elastic limit for stress corrosion specimens with rejection of an alloy if even one out of several specimens fails. The risk must be taken that an alloy which might have been just acceptable is given a bad reputation. In other directions, namely corrosion fatigue and erosion resistance, it is practically impossible to carry out proper mock-up tests on site. In wet conditions electrolytic corrosion between dissimilar metals must be expected, also crevice corrosion. Thus in the end the designer or the person who draws up the specification has to exercise judgment and this implies a cautious approach. The chemist or metallurgist often selects materials purely on general corrosion resistance and chooses the best, often 316 stainless steel which costs about 12 times the price of mild steel. This, though acceptable possibly in chemical plant where profitability is high, would make geothermal power plant so costly that it would be ruled out commercially. Some balanced compromise thus has to be reached. Price and corrosion resistance are not the only criteria: other required properties have to be considered, for instance austenitic stainless steels are particularly sensitive to stress corrosion cracking in the presence of chlorides. Moreover there are the following environments to be considered; an alloy found by test unsuited to one may be valuable in another:

(a) before separation where water and steam are both present and possibly rock particles;

(b) in steam pipes where initial contamination with salty bore water is gradually replaced by condensate;

(c) at the entry to the turbine where the steam is almost dry;

(d) at exhaust from the turbine where it is very wet, and impact erosion occurs;

(e) in the condenser where circulating water and air are present, gases may be concentrated and impact erosion may occur;

(f) in the gas extraction circuit where the most virulent mixtures of air, gas, water and vapour are found;

(g) in the circulating water circuit where gases may be released and flashing water and gases are discharged from drains.

So far only about 4 manufacturers have experience of geothermal turbines. Others have in the main shied away from the difficulties which their metallurgists, probably on the basis of hasty tests or judgment, tended to exaggerate. It appears that there is now a risk of a swing in the other direction with new manufacturers and ideas based on nuclear designs coming in with possibly erroneous notions on corrosion based on misunderstanding of published data or an excess of confidence derived from insufficient background knowledge.

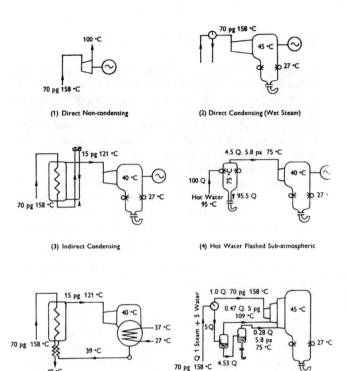

4. Types of turbines employed

The various possibilities are illustrated in Figure 1. The simplest process, applicable only where steam is available at above atmospheric pressure, is to put it through a turbine exhausting to atmosphere (Fig. 1 (1)). This is wasteful in that the heat drop available below atmospheric pressure is ignored, but may have to be done if the gas content of the steam is too high to permit of economical working under vacuum. Moreover in the initial development stage of a scheme when the field is being proven, it may be preferred for the small amount of power required locally to adopt a simple atmospheric exhaust turbine for the sake of economy in capital. Such plant up to 5 MW can be semi-portable, might be obtained quickly, and can be installed on temporary foundations. At a later stage the roughly 2- to 3-fold larger yield of power from a given steam supply may well justify the greater expense for permanent condensing turbines and circulating water works which are offset by the reduction in steam field costs per kilowatt. The heat drops utilizable in condensing and atmospheric exhaust cycles can be derived to a fair degree of accuracy from Figure 2 (discussed in Section 8).

Fig. 1. Various turbine arrangements.
p.a. = pressure in lb/in² absolute
p.g. = pressure in lb/in² gauge
Q = flow quantity.

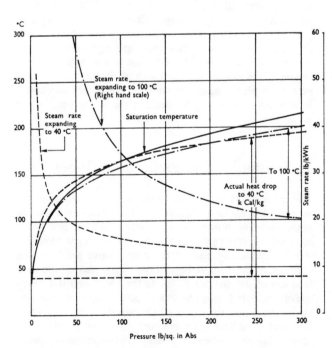

Fig. 2. Relation between pressure, saturation temperature and heat drop.

5. Condensing turbines

In all larger geothermal plants the raw steam is put through condensing turbines (Fig. 1 (2) (3) (4) (6))—a large fraction of the output—say $\frac{1}{2}$ to 2/3 being developed below atmospheric pressure. In some cases the steam may not be available above atmospheric pressure (e.g. where the source is boiling water at atmospheric pressure). In that case there is no choice but to run sub-atmospheric on flash steam or to resort to another fluid, as is dealt with in Section 19.

Condensing steam plant is more elaborate than that with atmospheric exhaust, needs more skilled attention, and involves more permanent structures, consideration of levels, and provision of circulating water and auxiliary plant.

6. Indirect system

In the early plant at Larderello the indirect system was adopted whereby the raw steam was taken through a heat exchanger and caused to re-boil its own condensate to yield pure steam at a lower pressure, the feed to the secondary (open) circuit being the (degassed) primary condensate. This avoided putting the gases in the raw steam through the turbine since the secondary steam was raised from degassed water. Thus (feared) corrosion from H_2S and CO_2 was avoided in the turbine though in some degree it was transferred to the heat exchanger. This practice is no longer adopted since it has been found that with suitable choice of constructional materials in the turbine, the corrosion problem is practically eliminated (in the absence of oxygen). The drop of temperature across the heat exchanger, say 25 °F, was also a severe handicap implying loss of say 20% of the potential power. Nevertheless this system might be revived if the gas content were so high as to render the gas extraction plant associated with the condenser too costly or cumbrous or if the chemicals in the steam were worth recovering (boric acid at Larderello).

7. Selection of pressure at turbines

The pressure at which the steam is to be utilised at the turbine depends in the first place on the pressure available at the wells. With normal well characteristics highest likely economic working pressure at the wellhead might be something approaching half of the closed valve pressure. An appropriate drop must be allowed in the steam pipe as a function of the distance to the farthest well and the steam velocity adopted. It is possible to do an economic exercise having obtained the wellhead characteristic relating pressure and yield and some pressure will show a maximum output. However there are usually many unknowns in such a study at the time it is carried out in that probably data are available from only a few wells which are not necessar-

ily representative of later ones. Their characteristics may differ. The likely life of the field and the rate of decay of pressure is not known. The cost per well, the cost of steam pipe as laid, the cost of turbines are all figures which have to be estimated and sometimes early estimates later prove wide of the mark. Hence in order to get ahead with the scheme the pressure has to be settled by judgment. The highest pressure so far employed is 180 lb/in² at Wairakei with multi-cylinder turbines. Many fields do not permit of the selection of so high a pressure, 100 lb/in² being more usual. This is about the limit set by the wetness implied at the exhaust, which can be accepted with single-cylinder expansion even to bad vacuum. When exhaust wetness would exceed about 13 or 14%, it is advisable to take the steam out and put it through a separator. Figure 2 shows that the heat drop available does not increase fast after a pressure of about 100 lb/in²g. Figure 3 deals with the issue in more detail for two particular vacua. A further complication in selection of pressure is that wells may be of two classes markedly differing in pressure either due to depth or other cause. To make proper use of both classes may thus demand two steam pipe systems operating at different pressures. This offers the advantage that boiling water at the higher pressure may be flashed into the lower pressure main and moreover high pressure wells, after falling below useful yields on the high pressure system, can be demoted to low pressure and given a new lease of life.

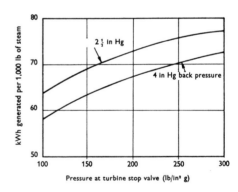

FIG. 3. Energy from saturated steam: condensing cycles for $2\frac{1}{2}$-in and 4-in Hg back pressure.

Because of thicker walls high pressure steam pipes, separators and valves are ostensibly more costly but this is offset by the lower specific volume of the steam (leading to smaller diameters) and the fact that each pound of steam has more capacity for work. Properly costed per kilowatt generated the high pressure parts of a steam turbine cost distinctly less than does the low pressure part. Indeed orthodox turbine price lists imply that the last row of large turbines, because of the extremely long blades, costs about 5 times the price per kW of the remaining part. The entire intermediate pressure cylinder from say 200 lb/in² down to 15 lb/in² also costs about half of the price per kW of the low pressure cylinder—from 15 lb/in² down to vacuum. Moreover the high pressure cylinder involves no condenser and the part of the station in which it is housed is cheap as requiring no deep basement.

As an alternative to separating the steam and the water in the field the proposal has been made since the very earliest times to transmit the steam and water together and to separate at the station but this has never been done except experimentally. It has the potential advantage that the drop in pressure enables a greater amount of steam to be flashed from the water. A drawback is that the large-scale discarding of the boiling water at atmospheric pressure is no easier at the station than in smaller amounts at the individual bores. Moreover having lost the benefit of the long steam pipe as a clean-up device, we require another. In a country subject to earthquakes, landslips, or unstable supports, the transmission of boiling water over long distances is also fraught with some danger.

8. Rough estimate of steam rate

In order to reach an idea of what output can be obtained from a given flow of saturated steam at a particular pressure, we require only the rough rule that each 1 degree F drop of temperature available results in a yield of 1 BTU/lb (Wood's rule). The validity of this is shown by Figure 2. The accuracy is $\pm 10\%$ which is something like the order of variation between turbines from the smallest to the largest sizes. There is nothing as simple as this rule applicable to superheated steam technology. For instance if steam is available at 100 lb/in²g (saturation temperature 338 °F) and is exhausted to atmosphere (nominal temperature 212 °F) the usable drop is about 126 BTU/lb. Since 3,412 BTU = 1 kWh, 1 BTU/sec = approximately 1 kW. Therefore 1 lb/sec of steam will yield approximately 126 kW. It is of course easy to make allowance for parasitic drops, e.g. since the exhaust will always have to be above atmospheric pressure, it might be taken as 16 lb/in²g, in which case the exhaust temperature will be 216 °F. If the steam were expanded to vacuum at a temperature of 100 °F, then the usable heat drop becomes 238 BTU and about 238 kW is the appropriate output for 1 lb/sec. Closer accuracy involves a more involved calculation yielding the type of result shown in Figure 3 drawn for $2\frac{1}{2}$-in Hg and 4-in Hg which is about the range of vacua usually attained by cooling towers in temperate and warm climates respectively.

9. Limits on unit size

The largest units operating are at Wairakei 'B' station being single-flow 30 MW machines running at 1,500 rpm with an inlet pressure of 50 lb/in² and back pressure of $1\frac{1}{2}$-in Hg. Their record has been substantially trouble-free. Several 50 MW two-flow 3,600 rpm machines are on order for the Geysers station to operate at 100 lb/in² and 4-in Hg back pressure (55 MW overload rating).

The limit size in any particular location may be set by the purchaser's specification in that heavier pieces of equipment may not be transportable to site. Another limitation is imposed indirectly by the allowable tip speed. This is selected with erosion in mind having regard to the wetness encountered, the vacuum and the confidence in erosion shields. Up to the time of Wairakei erosion shields were normally affixed by brazing and evidence both from Larderello and from test specimens indicated that brazing was attacked by H_2S in the presence of water. Erosion shields were accordingly avoided and the tip speed chosen appropriately low for soft blades (about 900 ft/sec). This moreover led to a low stressed design which was thought to be essential for avoidance of corrosion fatigue. Subsequent running has shown that judgment was correct in that erosion has not been severe and blade failures have not occurred in the low (or high) pressure turbines. However this does not prove what would have happened if higher tip speeds and therefore higher stresses had been permitted except that in intermediate pressure turbines with higher stresses some blade failures did occur attributable to resonance. At the Geysers on the other hand erosion shields were fitted on the first trial 12.5 MW turbine running with a higher tip speed and no trouble occurred (at least with the erosion shield fixing). This might be attributable to a different silver solder or mode of application. On the other hand the steam was less wet, partly due to the vacuum being much worse and also grease was injected as protection.

More recent developments such as electron beam TIG (tungsten inert gas) and even forge welding have made possible the fixing of stellite erosion shields without the use of brazing though it still may be prudent to adopt a moderate tip speed to ensure low stressing and hence lesser susceptibility to blade vibration failures. It is known that martensitic steels are particularly sensitive to H_2S attack. This means that the usual 12% Cr blade alloy must not be used in the hardened state. Most high strength materials are known to exhibit stress corrosion cracking and are generally regarded as treacherous in even a slightly corrosive environment. Hence it is unlikely that geothermal machines of comparable ratings to those of fossil or nuclear plants will be built. The corrosion problem will always call for more cautious stressing. Furthermore the erosion problem is more marked than in nuclear plant because the absence of boiler feed means there is no steam bled. Nuclear plants benefit from the fact that a good deal of water is carried away with bled steam. It is also worth remarking that the cascading of water from stage to stage in its flashing through crevices can cause severe cutting damage. This however can be repaired by deposition of 12% chrome by plasma torch or arc method. Experience shows that it is not necessary to line the entire inside surfaces of geothermal turbines with 12% Cr as in nuclear machines.

Further factors urging caution are the remoteness of many geothermal sites from the manufacturer's works, the high reliability which is demanded, and the very moderate scale on which the project is first initiated. Reliability is essential in that to show the best yield on the capital invested, a geothermal scheme must run at high load factor. To attain this means adoption of cautious design and the provision of spare turbine rotors.

10. Condensers

In all geothermal plants the condensers have been of the jet (spray) type in which direct contact is made between the condensing steam and the water. There are three good reasons for this. One is that the condenser cost bears a much larger proportion to that of the turbine in low pressure saturated steam cycles than on higher pressure superheated cycles as used in fuel fired plant. Secondly in fuel fired plant the condensate is required to be kept free of contamination because it is required for boiler feed whereas in geothermal plant no boiler feed is needed. Thirdly orthodox condensers traditionally use copper alloys for their tubes and copper alloys are basically undesirable in contact with geothermal steam because of risk of attack from H_2S. While at 'the Geysers' the condensers are constructed in 316 clad steel with stainless steel or aluminium piping for circulating water, at Wairakei and Larderello by contrast cast iron or mild steel have proved satisfactory (with coatings of various types). Internal surfaces need protection by stainless steel baffles where water jets would otherwise cause erosion-corrosion damage by impingement implying the removal of the protective patina or coating.

Initially jet condensers were built as cylindrical shapes and suspended beneath the turbine. Later models evolved into more nearly rectangular form. They were still squeezed into the confined space between the massive concrete columns supporting the turbine deck. An alternative idea derived from nuclear power stations is to build the jet condenser of such shape and form that it serves as a structure to carry the turbine. In this case the 'vapour' (gas) cooler which is used to reduce the volume of the mixture sucked off by the air pumps, is best incorporated in the main body.

The natural place for the condenser is immediately below the turbine and the simplest way of providing for the discharge of the water from the vacuum space is a barometric leg which requires to be something approaching 30 ft high. Thus the natural layout of the plant is with the turbine set some 40 ft above the level in the sump under the barometric tube (Fig. 4). Because this does not conform with normal power station ideas, alternative arrangements have in some cases been devised, for instance at 'the Geysers' and Matsukawa the condenser is placed high in the air outside the power station. This entails a long exhaust duct with a U-turn below the machine. The arrangement leads to various possible troubles associated with draining the U, also draining the pit in which it lies. There is also risk of water reaching the turbine blades with dis-

astrous consequences. It is concluded that where the ground formation permits, the condenser is better placed under the machine and with the barometric leg suitably offset where necessary reaching to a sump at a lower elevation. It is to be noted that where cooling towers are employed, two sets of pumps are ordinarily required, the first set to lift the water from the sump to the cooling tower, and the second set to lift water from the cooling tower basin to the condenser. Maintenance of desired water levels in sumps and basins may then call for automatic control. Occasionally as at Larderello the ground formation may permit of the cooling towers being placed on such a level as to enable the water to be lifted by the vacuum into the condenser instead of being pumped (Fig. 5).

11. Choice of vacuum

The appropriate choice of nominal vacuum depends in the first place on the source of cooling water, secondly economics, and thirdly convenience. The temperature of the cooling water, natural or recooled, is mainly dependent on the wet bulb. Because a great deal more cooling water is necessary per kilowatt than with orthodox steam plant, it is usual to keep down pump power by adopting a rather high range of cooling, 20-25 °F or even more. This means a rather poor vacuum. A poor vacuum is also advisable because of the large amounts of gas to be extracted. This helps to reduce the power or steam required for gas extraction from the condenser. A high range also helps the cooling tower—especially a natural draught tower where the hot water temperature largely dictates the draught obtained. It will be clear that the actual vacuum attained will vary with load and season. It may indeed be deliberately varied as a means of controlling load.

12. Cooling water

The required quantity of circulating water is arrived at by a rough rule that the latent heat of the steam at the condenser is of the order of 1,000 BTU/lb. Hence every pound of steam to be condensed requires 50 pounds of water with a temperature rise in the condenser of 20 °F (or pro rata). The fact that the steam will be wet at the exhaust is not important since the heat content above vacuum temperature of the water accompanying the steam is negligible. Where a more accurate calculating of cooling water requirement is called for it is necessary to know the heat consumption of the turbine. If for instance the performance is quoted as 15 lb/kWh of saturated steam at 150 lb/in² gauge and 2.5 inHg abs back pressure, then since the heat content of such steam is 1,196.8 BTU/lb and the condensate contains 76.6 BTU/lb.

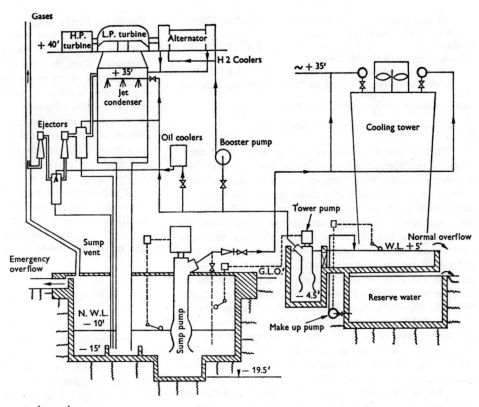

FIG. 4. Cooling water circuit diagram schematic.

FIG. 5. Diagram of circulating water system.

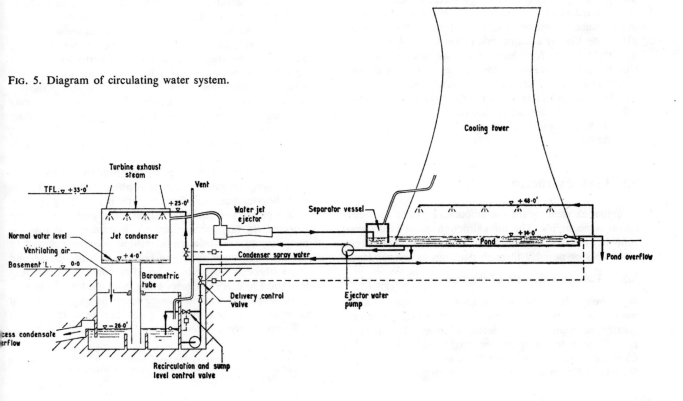

The heat rate is $15 \times 1,120 = 16,800$ BTU/kWh

We deduct from this the heat converted, say 3.450 BTU/kWh
(allowing for small outside losses)

And so obtain the heat rejected 13,350 BTU/kWh[1]

With a temperature rise of 20 °F in circulating water we require therefore $\dfrac{13,350}{20}$, or 667 lb/kWh.

This is 44.5 times the throttle steam flow.

While a source of natural cooling water is advantageous for geothermal stations as for orthodox stations (for instance a sizeable river, lake, or sea), a geothermal plant is unique in providing its own make up for cooling towers automatically. Hence in places without readily available cold circulating water, this can always be produced by cooling towers. Indeed there will ordinarily be a surplus of fresh water from the cooling towers because the heat carried off by the cooling tower is always in part (10 to 20%) in the form of sensible heat of air. Hence the remainder which is in the form of latent heat of steam requires only 80-90% of the steam condensed to be re-evaporated.

Cooling towers for geothermal stations are costly per kilowatt because of the much greater heat rejected per kWh. Natural draught towers may cost £6 to £7/kW instead of £2 in orthodox fuel fired stations. The water is pure condensate and can be concentrated to a high degree. It might be supposed that the content of H_2S and CO_2 would render the water strongly acid. This is not the case in practice. The CO_2 is largely scrubbed out in cascading through the fill. The H_2S is partly retained and partly oxydised to $SO_2 + H_2SO_4$ and S. Luckily geothermal steam usually contains traces of ammonia which, being highly soluble in cold water, is retained and builds up until it corrects the pH to the slightly alkaline side of neutrality. It combines with the HSO_4 to form ammonium sulphate and with the CO_2 to form ammonium bicarbonate. Some free sulphur is also produced which makes the water 'milky'.

Sulphate attack is likely to occur on concrete unless sulphate resisting cement is used (or equivalent pozzolana added). Where gases collect a resistant coating may be necessary.

13. Gas extraction

In orthodox steam plant the amount of gas in the boiler steam is negligible. The air extraction plant is designed primarily to extract air which has leaked in through flanges, shaft seals or possibly through packings at the condenser tubes. Though a formula of the form $m = \dfrac{W}{2,000}$, where W = weight of steam condensed/h and m = mass of air/h, has in the past been used, this is known to exaggerate very greatly the quantity of air which ordinarily has to be removed. As a token figure one might think of orthodox plant requiring air extraction plant to remove $\frac{1}{4}$ lb of air for every 1,000 lb of steam condensed. In geothermal plant the steam rarely contains less than $\frac{1}{3}$% of incondensible gases by weight and may contain as much as 4%. The leakage air may well be about the same proportion as orthodox plant, but the circulating water also brings in a large quota of air in solution commonly taken to be 3% by volume equivalent to 34 ppm in cold water. With a 50 times ratio of water to steam even this small amount of dissolved air in the circulating water implies 0.17% of steam flow. Hence it will readily be seen that the air or gas extraction plant must be much larger with geothermal plant using jet condensers than in orthodox plant. Moreover the gas pumped out is a mixture of air, CO_2 and H_2S. This is a higlhy corrosive combination entailing rubber lining, fibre glass, or stainless steel piping if periodic patching (and replacement) of mild steel is not acceptable. Luckily some of the gases are entrained in the water going down the barometric tube.

14. Types of air extraction plant

The most favoured method for many years in orthodox plant was the steam jet ejector. This replaced about 1906 the reciprocating air pump which had been improved by using a steam jet in the form of a Parsons vacuum augmenter enabling it to reach a higher vacuum. When the steam jet ejector took over it was claimed to be efficient because the heat in the jet steam was recovered by warming the condensate. In fact this claim was spurious because by 1906 feed heating by bled steam had already been introduced and thereafter only a small credit could be given to the heat recovered from the jet ejectors. Nevertheless because available, they have remained fashionable even though their efficiency is very poor, not better than 5%. For large sizes they may be noisy. A better efficiency is obtained in orthodox plant by using water jet ejectors which, though they work on the same momentum principle, do not entail the same rejection of latent heat. However they require large amounts of water for geothermal duty and cost more than steam jets (shown in Fig. 5). Economic choice is hence somewhat dependent on the value that is put on the steam.

Consideration also has to be given to the means of discharging the gases to the atmosphere. H_2S and CO_2 are both heavier than air. Consequently if cold they may lie in low places and could be dangerous as neither is readily detected (in high concentration). Accordingly it may be necessary to over-ride considerations of efficiency and to use steam ejectors since the hot steam from the second stage (there is no sense in an after condenser) provides levitation to carry the gases away from the top of the vent pipe. A natural draught cooling tower forms a convenient chimney to ensure that the gases can be discharged at sufficient height.

---

1. This is the total of heat rejected from condenser, oil cooler, and generator cooler. If the condenser alone is considered, a deduction of 3,550 to 3,600 is to be used above.

15. Compressors

Even more efficient than the water jet ejector is a centrifugal compressor. Where the gas quantities to be handled justify it these may be used (efficiency say 60%). They have run successfully for many years at Larderello in multi-cylinder form with spray intercooling—total ratio about 10/1. Rating is 900 kW, either turbine or motor driven. Corrosion and erosion troubles have been absent despite the wet and corrosive gases. Water-ring pumps with eccentric casings which behave as compressors are now used in orthodox plant often in combination with a first stage steam jet. This is the most effective method on overall cost considerations. They have not been used on geothermal plant where the capital cost might seem high. A possibility not so far utilised is that of operating the ejectors on boiling water. The notion that a high vacuum cannot so be obtained is erroneous (Frenzl, 1958).

16. General precautions on gases

Experience shows that safety precautions such as avoiding unventilated pits in basements where gases can collect and taking sufficient care over venting gases away at a sufficient height, also dealing adequately with miscellaneous drains which all carry gases, are not only advisable from human safety and comfort aspects. They also help to avoid failures of minor electrical apparatus such as instruments and relays where fine copper or resistance wires (even silver) are readily attacked by traces of H_2S. Such troubles are not readily avoided by specification since such minor items are supplied by subcontractors as standard equipment without consideration of the specification. An outdoor plant is less troublesome in these respects than a normal power house. Copper commutators of exciters and copper 'pigtails' of brushes have also shown corrosion trouble. These can be overcome in various ways, the most radical of which now available is to eliminate commutators by use of ac or static exciters.

It is worth mentioning that the most severe corrosion damage to geothermal steam turbines has occurred during periods when they were standing still. This is because valves are rarely steam tight and a small amount of steam leaking into a turbine in the presence of air will do severe damage because the gases are concentrated in the condensate and water line attack results. Hence when geothermal plant is not in operation precautions must be taken to ensure tight steam shut off by provision of two valves in series with a vent to atmosphere between, and preferably also by drying out the turbines with hot air after shut down.

17. Regulation of geothermal power stations

Geothermal plant is best suited to base load duty. Therefore it ideally should be only a supplement to other plant on an interconnected system. Other plant, assumed to be of higher running cost, takes the peaks. The reason for restricting the geothermal plant to base load is not only that the cost of production is low because it has no overt running cost (equivalent to fuel) and the incidence of capital cost is reduced if spread over as large a number of kilowatt hours as possible. There are indeed physical limitations on operating geothermal plant at other than base load. The steam source must be regarded as somewhat equivalent to a flowing river and the geothermal plant is therefore similar to a run of river hydraulic station. The analogy is not complete because while the river flow is decided by nature and varies with the rainfall but has an infinite life, the amount of steam can in fact be adjusted by bringing on or shutting off wells but the field has a (likely) finite life. However cutting off and putting on wells is clearly something which cannot be done quickly or remotely from the power station. Hence the most convenient way to operate a geothermal power station is to let it produce a suitable constant load allocated to it in advance. If desired this allocated load may be somewhat below its load capability so that the turbines run normally at say 95% of full output and hence with their throttle valves slightly closed. Thus on fall of system frequency the geothermal plant will respond by opening the throttle valves and increasing load up to the maximum swallowing capacity of the turbines. This is possible by pulling down the pressure previously absorbed in the throttle valves. In the previous condition there may have been a small blow-off from relief valves (equivalent to a small spill over a weir at a hydraulic station) though this is by no means essential or desirable and indeed for best economy the load should be regulated to avoid it. The issue is of course bound up with the question of the amount of spinning reserve called for on the system to meet sudden load demands and the suitability of other plant for meeting this duty.

On throwing off load such as may occur on tripping out of the transmission lines between the geothermal power station and load, the turbine governor valves will close and consequently the steam pressure before them will rise. The relief valves must then come into operation to limit the pressure to a safe value and discharge the excess of the steam yielded by the bores. At times also when part of the plant is out of operation, excess steam may have to be blown off either through relief valves or better through hand controlled discharge orifices until such times as wells are shut off. Shutting wells off will clearly be undesirable for a short period as involving too much work for a possibly small team of field staff. Furthermore some wells if shut off are not self-restarting but have to be induced to flow. In that case it is clearly more convenient to blow off steam except in the case of a long shut down. Again the attitude is influenced by how long the steam field is expected to last. If the life is thought to be infinite or cannot be estimated, then there is no great interest in steam conservation.

18. Utilisation of hot water

As mentioned earlier it is usual where wet steam is obtained to separate the steam from the boiling water at the wellhead and to discard the boiling water there. The consequence is that a great deal of heat of moderate potential is thrown away (possibly as much as is contained in the steam). Because part of the boiling water separated at a pressure above atmospheric flashes on reaching the atmosphere, there is also a great deal of visible vapour produced and associated noise. Discarding boiling water, from which power could be developed, may appear inherently wasteful but the decision is governed by economics. Clearly to utilize the water at the wellhead would entail a multiplicity of small power plants. To make good use of the low pressure flash steam obtainable would entail condensing working which is out of the question on grounds of complication. An alternative is to resort to a heat exchanger and employ a refrigerant in a closed cycle which in principle might use an air-cooled condenser so avoiding need for cooling water. This again is too complicated for small scale, all small scale generation being relatively costly in capital and in attendance.

A third possibility is a boiling water turbine which has not been tried yet. The isentropic heat drop available in boiling water at say 380 °F 181 lb/in²g (hw = 353.6 s = 0.5413) above atmospheric pressure can readily be calculated from steam tables (hw = 180.2 s = 0.3122 for water at 212 °F) as:

$$h_1 - h_2 - T_1(s_2 - s_1)$$
$$= 353.6 - 180.2 - (212 + 460)\ 0.2291$$
$$= 19 \text{ BTU/lb.}$$

The heat energy can be converted to head by Joule's equivalent 778 ft = 1 BTU/lb. Therefore the above 19 BTU/lb is equivalent to 14,800 ft which is a formidable head by hydraulic standards demanding a new form of turbine unless the drop is taken in two stages. It then comes within the range of the highest hydraulic heads employed but there will be flash steam generated in the first stage calling for a steam turbine to utilize it, in parallel with the water turbine.

If we reject the notion of utilising the energy in the boiling water in the field, the alternative is then to pipe it to the power station where it can in principle be utilized by any of the above methods. The simplest is to flash down to about atmospheric pressure to yield low pressure steam which may be admitted into the main turbines. This has been done at Wairakei though it entailed development of rather large steam scrubbers to avoid contamination with chlorides carried over in spray. The plant was put out of commission however when the wells which served it went dry and the water pipe was converted to use as a steam pipe.

In all such cases it is difficult to decide until a technique has been established whether the capital cost can be justified. There are commonly several available alternatives. Economic choice is based on 'skimming the cream', i.e. utilising only that portion of the available energy that can be developed cheaply with reasonable certainty of continuing for sufficient time to justify the initial capital investment.

The notion has many times been put forward that the hot water should be got rid of by returning it into the earth. The objection to this is that while the water still contains heat, the level of this heat is de-graded in relation to that at its point of origin and hence its return to the source could only result in lowering the source temperature and thus helping to destroy the field. However return to earth may be the only choice if the Catchment Authority will not allow water to be turned into streams via surface drainage.

Consideration has also been given in some cases, e.g. in Iceland, to utilisation of the boiling water elsewhere. Two possibilities discussed at Hveragerdi were to pump it to Reykjavik where the town was already heated with geothermal water, or to use it in local greenhouses which are already dependent on geothermal water. The former plan was found to be too expensive because of distance and high ground intervening. The latter could not utilise more than a small fraction of the heat available. Salt production by distillation of sea water was also considered.

19. Consideration of alternative fluids to steam

So far with two exceptions geothermal plants have used steam as the working fluid. The exceptions are a 300 kW plant on the island of Ischia, Italy, in 1948 using ethyl chloride and a recent 500 kW plant in Kamchatka using Freon 12. The ostensible attraction of such plant is the possibility of utilising hot water at a temperature below 212 °F (100 °C) without going sub-atmospheric in pressure. The refrigerant will have a higher vapour pressure and thus the turbine will be more compact by virtue of the higher density of the vapour. The plant can be self-starting. Risk of corrosion in the turbine is also avoided as is wetness loss with some refrigerants. Corrosion risk is of course transferred to the heat exchanger and (if geothermal water is used for cooling) to the condenser which now has to be of the surface type (Fig. 1 (5)) (there may be no other water).

Because refrigerants commonly show a high specific heat in relation to latent heat of vapour, more heat might apparently be extracted from the hot water than with the single flash process. Some writers (Hansen, 1961; Pradhan, 1970) have claimed that theoretically a refrigerant used above its critical pressure will yield much higher output than does water. In practice this does not seem to be true. Moreover the necessary heat exchanger incurs capital cost while the temperature drop across it reduces the potential and therefore the heat drop available. Secondly the surface condenser required is dearer than a jet condenser. The feed pump commonly takes considerable power. These are serious handicaps which offset the reduction in the turbine dimensions. The issue merits the following detail discussion:

19.1 COMPARISON FOR SOURCE OF HEAT AS BOILING WATER AT 450 °F

For the general case it is convenient to regard the source heat merely as sensible heat in boiling water, either under pressure or at atmospheric pressure. The simplest plan if the water is under pressure is to flash it down to a lower pressure, above or below atmospheric and expand the resulting flash steam down to vacuum, condensing in a jet condenser. The second (Fig. 1 (5)) is to exchange the heat in the water with a refrigerant boiled at a suitably chosen temperature, to expand through a turbine to a surface condenser, recycling through a feed pump in a closed circuit. The main considerations are how much power will be produced and at what capital cost. The vacuum temperature is taken in each case at 100 °F (see later discussion).

The exercise as regards output may be done on isentropic calculations which are simple and good enough for a first assessment taking boiling water at 450 °F as the source and an assumed 'nip' temperature drop across the heat exchanger of 20 °F. Comparison is made first with flashed steam obtained at the optimum of 270 °F in single flash and also with double flash, the criterion being the useful power extracted from one pound of source water (quoted in BTU), which for water comes out as about 49 (Fig. 6) for single flash and 61.4 for double. We have a choice of whether or not to use superheat in the refrigerant. The curves of Figure 6 indicate that superheating in no case pays for the reason that for a fixed quantity of heat available in boiling water the gain in refrigerant specific heat drop by superheat is less than the resulting reduction of mass flow. Feed heating by bled refrigerant also has no apparent advantage because there is always sufficient low grade heat in the waste hot water to carry out the feed heating in the heat exchanger. Indeed the refrigerant can extract more heat from the water only because it is heated from vacuum temperature.

Some refrigerants yield less power than does single flash water but some might produce more. R113, though little used, offers the initial potential advantage of some 9% bigger theoretical output than does the (single) flashing process. The hot water thus gives up more heat and is rejected at a lower temperature than the 270 °F with single flash. The comparison is rather sensitive to the 'nip' temperature across the exchanger (Fig. 7) e.g. if this is as high as 40 °F then no refrigerant used subcritical matches water. The assumption also has a profound effect on the cost of the heat exchanger which in principle for 20 °F will cost twice as much as for 40 °F mean temperature difference. Allowance is made for the ideal feed pump power which with water is absent but in refrigerants may be quite large

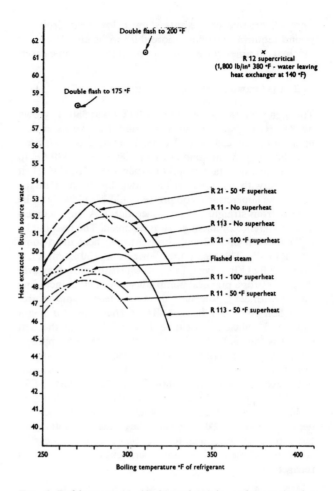

Fig. 6. Refrigerants 11, 12, 21 and 113 in cycles condensing at 100 °F. Heat extractable from source water at 450 °F by heat exchange and isentropic expansion to 100 °F.

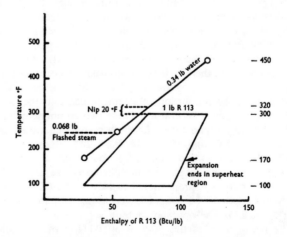

Fig. 7. Temperature-enthalpy diagram for 'R 113' boiling at 300 °F: no superheat. (For comparison the quantity of flashed steam produced at 250 °F is shown.)

since all organic refrigerants show a low heat drop per pound requiring therefore more fluid to be circulated—as much as 10 times corresponding steam flow in some cases.

19.2 SUPERCRITICAL USE OF A REFRIGERANT (R12)

The high yield in Figure 6 of 61.4 BTU/lb at 380 °F shown by R12 (C Cl_2F_2) is attained by using it above its critical temperature of 234 °F and critical pressure of 597/lb/in² abs choosing initial conditions of 1,800 lb/in² 380 °F to produce just dry state at end of expansion to 100 °F. Heat exchange with the water is facilitated by the heat intake line in the supercritical compressed fluid being likely to be not far from straight (no published data). A 'nip' temperature difference of about 40 °F is assumed. The heat intake is 75 BTU/lb of R12 and the isentropic heat drop some 18.7 BTU/lb. The feed pump however at 100% efficiency takes 3.9 BTU/lb leaving 14.8 nett on isentropics. However if we allow for a practical pump efficiency of 70% including motor with 25% pressure drop the feed pump power deduction rises to 7 BTU/lb. The turbo-alternator might show 85% efficiency, yielding 15.9 BTU/lb and the nett output is thus 8.9 BTU/lb. The exchange ratio is approximately 4.1 lb of R12/lb of water. Hence there is prospect of coming out with a practical 36.9 BTU/lb of water. This is however no better than obtainable by double flash (37.5).

The capital cost of supercritical R12 plant is likely to be high in view of the high working pressure for which the heat exchanger (1,800 lb/in²) and the condenser (130 lb/in²) have to be designed. The Russian plant referred to earlier seems to represent a full scale experiment in the use of this refrigerant.

19.3 FIRST FLASH OPTIMIZATION

Initially 250 °F was selected arbitrarily implying a pressure in the flash vessel a little above atmosphere as might be preferred (at least at full load). Furthermore 250 °F is about as high a temperature as many refrigerants are capable of utilising without instability, and indeed is above the temperature at which they are normally used. Hence it is in the region where there is little or no actual operating experience with them. The optimum flash temperature was found by the following method. The turbine output is the product of the proportion of steam flashed and the heat drop to vacuum temperature. These two components are covered by the two rules of thumb mentioned earlier in the chapter. Taking 250 °F we have 20% flash × 150 BTU/lb = 30 BTU output per pound of initial fluid. If we generalise by taking the flash temperature as X °F, then the output is proportional to $(450 - X)(X - 100)$ which is seen to be quadratic vertex up with maximum at X = 275. The optimum output is then 30.6 BTU/lb. The relationship of this 30.6 or 30 BTU/lb of available water to the isentropic yield quoted above of 48 to 49 is explained by the latter being a theoretical figure and 30 a practical one after making allowance for turbine and

generator inefficiency and also leaving loss. It is to be remarked that the heat contained (above 100 °F sink temperature) in boiling water at 450 °F is 362 BTU/lb. Hence the efficiency of utilization is only 8.3% in the single flash process. However the isentropic head available above sink is only 83 BTU/lb so we achieve 36.6% of the theoretical.

19.4 SECOND FLASH OPTIMIZATION

It might appear that all we have to do is to choose the second flash temperature at 175 °F by the same process having fixed the first flash at 250 °F. This however is not the case; the combination must be reconsidered and in fact the maximum combination output occurs with a first flash temperature of 310 °F and 200 °F for the second, though output is not sensitive over some range. These give the high output on isentropics of 61.4 BTU/lb shown in Figure 6 which happens to equal supercritical R12.

19.5 LOWER TEMPERATURE WATER

It might be thought that the high source temperature assumed is particularly favourable to water. The exercise has however been carried out also for lower source temperatures with much the same result. With reduced temperature scope available parasitic pressure and temperature drops in the closed circuit become more serious and the capital cost of the heat exchanger accentuated because more heat has to be transferred per kilowatt produced.

19.6 CONDENSERS

It has been assumed in all the above that the same condensing temperature of 100 °F is attainable in either case even though in principle for a given cooling water temperature the direct contact process in the jet condenser should result in a closer approach. This however tends to be offset by the presence of gases in the steam and accordingly the rough assumption of equality of vacuum temperature is reasonable. A tubular condenser for steam would however be expected to cost roughly double the price of a jet condenser (even overlooking the corrosion aspect). A refrigerant condenser may cost even more because the vapour pressure is higher calling for thicker walls, and the heat transfer rate on the refrigerant side is generally lower than that in condensing steam. To suit the pressure in the refrigerant it might be necessary to condense it within the tubes so that there is then likely to be a need for extended surface on the inside of the tubes. Very little information is available on the costs of such equipment and we can do no more in the general case than deduce that even though the turbine might be very much smaller than the equivalent steam turbine, any gain in that direction is likely to be about offset by the cost of the heat exhangers and inventory for working fluid.

19.7 SUB-ATMOSPHERIC OPERATION

From a practical point of view there is nothing against sub-atmospheric flash except the need to seal against inleakage of air and inability to start unless vacuum is raised by external power. These considerations apply also to single flash for lower source temperatures. Indeed as hinted above even the first flash will result in sub-atmospheric pressure in the turbine at light load. The only thing that has to be overcome in adopting a second flash is a certain reluctance on the part of the steam turbine maker to build a machine with a low pressure admission belt which makes it different from his normal run. However single-stage turbines (Bauer-Wach) have been built in the past for operating on the exhaust of condensing steam engines. Valves for controlling the admission of the low pressure steam tend to be rather large. The maker is inclined to think of the extra swallowing capacity in the low pressure stages of the turbine as being costly to manufacture. Nevertheless the last stage at possibly 5 times the average price of the remainder is generally easily justified economically because the extra power produced is obtained without corresponding extra steam production costs. In orthodox plant it does not incur a cost for fuel or for boilers. Likewise in geothermal plant it does not entail a cost for wells or pipework provided that the water from which the flash steam is obtained has already been conveyed to the power station.

19.8 GENERAL ASSESSMENT

To summarise, it appears that the claims which are often made that some other fluid than water is more suitable for generating power in the low pressure region, do not seem to be proven. While it is conceded that theoretically a refrigerant can promise more output than does water in the single flash process, it appears that double flash can about match anything that refrigerants can hope to attain and is eminently more practical. Clearly where the heat is available in the form of steam or boiling water initially, there is the great virtue of simplicity in utilising the steam directly or deriving steam by the flashing process. In all cases so far investigated it seems likely to be cheaper to do this than to exchange heat into another fluid. We have however reliable cost data only for steam and until the exponents of some other fluid publish their figures we remain without actual cost data for the alternative.

One possible advantage of a refrigerant is ability to provide independent start-up which is not obtainable readily with sub-atmospheric steam though might be done in some cases by throttling a well to give higher pressure steam or boiling water for the ejectors. Against this may be set the inability of a refrigerant turbine to provide its own pure make up for the water cooling circuit thus attaining self-sufficiency in that direction. Moreover many other types of plant, e.g. gas turbines, diesels, and modern steam plant are also incapable of a 'black start' except by reliance on some form of stored energy; indeed geothermal plant and hydro are almost unique in offering independent start-up.

Favourable conditions for a refrigerant are:
(a) where the steam contains a high content (say 10% of incondensible gases. If a secondary cycle is required for steam then it has no initial advantage over another fluid;
(b) where, as in the U.S.S.R., the vacuum temperature in winter might be so low that freezing risks occur with water and because of the very high specific volume the annulus area required with steam becomes inconveniently large.

In the beginning it was somewhat difficult to persuade manufacturers that power from geothermal steam was a sound proposition even though the steam was not very different from that which they were familiar with in existing steam turbines. One can only suppose that there would be even more difficulty in persuading manufacturers and others of the value of developing a peculiar turbine to utilise a refrigerant, particularly as there is a great variety of refrigerants. The turbine design would be quite different from that of steam turbines and there would be no assurance that a development project would lead to continuing business, however interesting academically.

Bibliography

DORSEY, N. E. 1940. *Properties of ordinary water substances.* New York, Reinhold.

FRENZL, O. 1958. Souffleries intermittentes à trompe d'induction par eau chaude, *Docaero*, no. 51, July 1958, p. 3. Paris, Centre de documentation de l'armement.

HALDANE, T. G. N.; WOOD, B.; ARMSTEAD, H. C. H. 1958. The development of geothermal power generation. *World Pwr Conf., Montreal 1958* (Paper 21 C/1).

HANSEN, A. 1961. Thermal cycles for geothermal sites and turbine installation at the Geysers Power Plant, California. *Proc. U.N. Conf. on new sources of energy, Rome 1961.* New York, United Nations (Paper 35/G/41).

PRADHAN, A. V. 1970. *Supercritical cycle gas turbine power plant utilizing geothermal water energy.* New York, American Society of Mechanical engineers (Paper presented at the A.S.M.E. Gas Turbine Conference & Products Show, Brussels, Belgium, 1970—70-GT-60).

WOOD, B. 1960. Wetness in steam cycles. *Proc. Instn Mech. Engrs, Lond.,* vol. 174, no. 14, p. 491.

——. 1965-66. Wetness in steam turbines—a general progress survey. *Proc. Instn Mech. Engrs, Lond.,* vol. 180, part 30, p. 1.

——. 1969-70. Alternative fluids for power generation (April 1970). *Proc. Instn Mech. Engrs, Lond.,* vol. 184, Part 1.

Geothermal district heating

Sveinn S. Einarsson

Managing Director of 'Vermir' H/F,
Research Engineers and Geophysicists,
Chairman of Board of Directors of the
Icelandic Institute of Industrial Research
and Development, Reykjavik (Iceland)

1. Introduction

The optimum ambient temperature for human comfort is
in the range of $+$ 15-22 ºC, depending on physical effort
and to a certain degree on environmental factors such as
relative humidity, air movement, exposure to radiation
etc. The temperature of the natural environment in the
inhabited parts of the earth varies, however, within the
wide range of say $-$ 35 to $+$ 45 ºC. Few parts of the earth
offer the optimum temperature except for a relatively short
span of the yearly cycle.

Clothing, houses and other shelters give protection
against exposure to extreme cold and heat, but are insufficient for full comfort, unless the confined spaces are
heated or cooled to the desired temperature. Artificial
heating or cooling of houses requires expenditure of energy,
which is supplied primarily by fossil fuels.

However fuels are costly, the reserves (while enormous)
will ultimately be depleted, and their use pollutes the environment on a scale that is now becoming a real concern
in many towns and metropolitan areas of the world.

An alternative source of energy that is eminently
suited for this purpose is the geothermal energy which has
been receiving increased attention in a number of countries
in recent years.

Prospecting for geothermal energy is still only beginning, and accordingly estimates of the global reserves are
bound to be uncertain.

The known reserves are nevertheless tremendous and
growing every year as prospecting is continued. This is
perhaps best illustrated by random examples.

Measurements of terrestrial heat flow in Hungary
were carried out in 1954 and led to the discovery of reserves
of 4,000 km³ of hot water (60-200 ºC) stored under the
Hungarian plains. The recoverable heat has been estimated
by Boldizsar (1970) at 2.3×10^{19} cal which is about 50%
of the calorific value of the known petroleum reserves of
the world. Important discoveries have been made in recent
years in other parts of the world. It is thus claimed by

Tikhonov *et al.* (1970) that thermal waters available for
economic exploitation are found at depth in 50-60% of
the territories of the U.S.S.R. Enormous hyperthermal
areas are known in regions of recent or active volcanism
in Europe, Africa and a number of countries surrounding
the Pacific Ocean. The potential of the hydrothermal systems represents only a fraction of the energy reserves stored
in the deep-lying rock formations of the earth's crust,
provided that a suitable technology of extraction could be
developed.

The utilization of geothermal energy presents as a
rule minimal pollution problems, much less than those
inherent to the use of fossil fuels. Most important, however, is the fact that the production cost per unit of energy
is lower for geothermal energy than that of most if not all
other available sources of energy.

Where geothermal energy is abundant it is in fact
sufficiently economical that artificial microclimate can be
created in relatively large sheltered spaces, say over scores
or hundreds of hectares. This is of greatest importance for
agriculture in countries with cold climates. But another
obvious use for geothermal energy is the heating of houses
in cold climates and cooling of houses in hot climates for
increased human comfort.

The present paper will report on developments that
have taken place in utilizing geothermal energy for house
heating in various parts of the world, describe the technology employed and discuss some pertinent economic aspects.

2. Historical notes

(a) ICELAND

The use of geothermal energy for large-scale heating of
houses was pioneered in Iceland, which is natural in as
much as the country has a cold temperate climate requiring
the heating of houses for about 330-340 days of a year.
No fossil fuel sources are available in the country except
limited reserves of peat and lignite. Hyperthermal areas

are however abundant, as manifested by a multitude of hot springs and a number of major geothermal steam fields.

The first attempts at heating single farm houses were made in the beginning of the 20th century, and around 1925 hot spring water was used for the first hot houses for vegetables (Bodvarsson et al., 1964; Zoëga et al., 1970).

A major step forward was made when the first bore-holes for hot water were drilled in the vicinity of Reykjavik in 1928, yielding 14 l/sec of 87 °C hot water. A 3 km long pipeline to the city was constructed, feeding a pilot district heating scheme comprising 70 houses, an enclosed swimming pool, an open-air swimming pool, and a public school house (Sigurdsson, 1964).

This venture was quite successful, and as a consequence more extensive drilling for hot water was started in 1933 at the Reykir thermal spring area some 18 km outside the city. This yielded about 200 l/sec of 86 °C hot water, and in the years 1939-1943 a new district heating system was constructed, serving 2,300 houses with about 30,000 people and all public buildings of the town. The system included an 18 km long transmission pipeline, pumping stations, 8,000 m³ water storage tanks and underground distribution network in the streets of the town. This system was commissioned on 1 December 1943 and is still operating quite satisfactorily.

An extension of the system was built in 1949-50 after additional drilling had been carried out 3 km north of Reykir. A total of 72 boreholes were drilled in these two spring areas, producing a total of 330 l/sec of 86 °C water. The boreholes vary in depth from about 300 m to 770 m.

The Reykir fields have now been producing continuously for about 35 years, and the total production over this period is of the order of 300 million m³ of water without any drop in temperature or decrease of the production rate.

A geoscientific survey of the area within the limits of the city of Reykjavik was resumed in 1954 (Bodvarsson and Palmason, 1964) and delineated a thermal area where deep drilling (700-2,200 m) was started in 1958. The steady yield of his field is now 300 l/sec of 128 °C water, and another thermal area more recently discovered and developed on the eastern periphery of the city yields about 165 l/sec water at 105 °C (Zoëga et al., 1970).

The district heating system has been greatly expanded on the basis of these new thermal sources and 8,700 houses with 72,000 inhabitants were connected to the system by the end of 1969. This represents about 87% of the houses in the city, and the goal of connecting new residential quarters to the heating system as soon as they are built is nearing realization.

The city of Reykjavik has developed, owned and operated the systems from the very beginning. It has been highly sucessful, technically and economically, and is perhaps more highly regarded by the inhabitants than any other public service. This has contributed to the general concept in Iceland that, where thermal resources are available, a district heating system is a necessary and normal public service, just as much as a cold water supply, an electricity supply or a sewer system.

Developments in other parts of Iceland have been somewhat slower than in the capital, mostly because of lack of suitable financing resources for the rather capital-intensive geothermal installations.

The town of Olafsfjördur in northern Iceland with 1,000 inhabitants built its first district heating system in 1944 using hot water of only 48 °C produced by drilling in a thermal area about 3 km away. Deep drilling in 1961 yielded additional water at 56 °C. The system comprises all the houses of the town, and it has been quite successful.

The town of Selfoss (2,200 inhabitants) in southern Iceland has a district heating system which was built and operated by a local cooperative society from 1948 until 1969, when it was sold to the township. It heats the whole town comprising 154,000 m³ of flats and 75,000 m³ of public, commercial or industrial buildings. The water is supplied from boreholes, 1.5 km outside the town and the production is 80 l/sec water at 80 °C.

The village of Hveragerdi in southern Iceland with 820 inhabitants and a balneo-therapeutic institution for 140 patients, is one of the main hot house centres for growing flowers and vegetables. A district heating system serving all houses in the village and supplying heat to 30,000 m² of hot houses has been in operation since 1953. This is the only district heating system in Iceland that uses a high-temperature field (180 °C).

The town of Saudárkrókur in northern Iceland (2,000 inhabitants) has operated a district heating system for about 15 years, utilizing water at 70 °C from boreholes in the vicinity.

Since about 1930 elementary and secondary boarding schools in the rural areas of Iceland have whenever possible been sited at locations where geothermal energy is available. In these centres the school buildings and living quarters for the pupils and staff are geothermally heated. They are also as a rule equipped with a swimming pool, and are self-supplying with vegetables (tomatoes, cucumbers, cauliflowers, etc.) grown in their own hot houses. There are now many such schools in various parts of the country, and quite often they are used as tourist hotels during the summer holidays. Quite often these centres have formed the nuclei of new service communities in the rural areas.

Several new district heating schemes have been studied, and some are now under construction. A scheme for supplying the satellite towns of Reykjavik with geothermal heat from the neighbouring high-temperature areas has been studied. Feasibility studies have been made for district heating systems for Akureyri (10,000 inhabitants), Siglufjördur (2,500 inhabitants), Akranes (4,500 inhabitants), Húsavik (2,000 inhabitants), Keflavik International Airport and the neighbouring communities of Njardvik and Keflavik (about 10,000 inhabitants), Dalvik (1,000 inhabitants), Hrísey (300 inhabitants), Egilstadir (800 inhabitants), Reynihlíd at Lake Myvatn (200 inhabitants and some tourist hotels). Several of these projects are already being executed or are being seriously considered.

At the end of 1969 a total of almost 80,000 people, or 40% of the population of Iceland (200,000), lived in houses heated with geothermal energy. On the basis of the present geographical distribution, the author has estimated elsewhere that 60-70% of the population could be supplied with direct geothermal heating for their homes. Figure 1 shows the gross production of geothermal energy in Iceland since 1960 with a forecast of the growth until 1975 (Palmason and Zoëga, 1970).

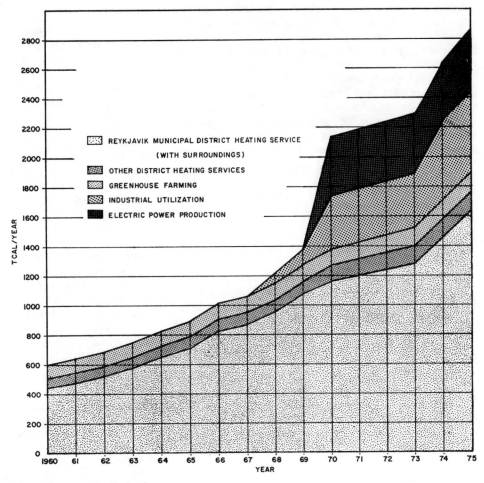

FIG. 1. Gross production of geothermal energy in Iceland (above 40 °C).

(b) HUNGARY

Following the discovery of the vast geothermal reserves of Hungary about 15 years ago, considerable interest was aroused for utilizing the energy for domestic heating. Plans were studied for constructing a geothermal district heating system for the city of Szeged around 1962. However, the drilling for hot water for this project led to the discovery of Hungary's richest oil and gas field, which resulted in the heating system being converted from geothermal energy to gas. Subsequent utilization of geothermal energy in Hungary has concentrated on agricultural applications (hot houses and animal husbandry), which have expanded extremely rapidly (400,000 m² hot houses by the end of 1969, expected to be doubled in 1970) (Boldizsar, 1970).

The total volume of buildings heated with geothermal energy by mid 1969 is reported as follows (Boldizsar: private communication):

Szeged 1,200 flats	120,000 m³
Szeged University clinics	106,000 ”
Hódmezóvásárhely factory	160,000 ”
Hódmezóvásárhely Hospital	12,000 ”
Makó Hospital	80,000 ”
Total	478,000 m³

The thermal waters of Hungary produced by drilling to a depth of about 2,000 m have a temperature of about 90 °C and the yield per well is of the order of 15-30 l/sec on the average.

(c) JAPAN

The traditional use of thermal springs in Japan, dating back through centuries, is for recreation and health (balneological and therapeutic uses). It is thus estimated that 100-150 million visitors enjoy annually the recreational facilities offered by the thermal springs of Japan (Komagata et al., 1970).

Thermal waters have been used for hot houses for flowers since 1916, and hotels and other facilities of the recreational centres are now heated with geothermal waters.

The construction of geothermal district heating schemes involving comparatively long-distance transmission of thermal waters has been reported (Mashiko and Hirano, 1970) in the following localities:

An 11.5 km long transmission line carrying 14 l/sec of 70 °C water from the Sarukura springs to the town of Towata, was constructed in 1963.

A 12 km long transmission line in the Okawa area was built in 1963. It carries about 22 l/sec of 70 °C water and supplies 3,000 houses with heat on an area of 260 hectares.

A district heating system was built for the Ukiyama area in 1965, comprising 900 houses on 100 ha. The system is equipped with a boiler plant that can heat the water to 55 °C. The total length of pipe is 12 km.

A district heating system for the city of Aomori was constructed in 1966-67. It supplies 140 houses, including 34 hotels, and the population is 3,600. The water is supplied from the Asamushi hot spring area at a flow rate of 22 l/sec and 60 °C. The natural springs have temperatures in the range of 40-70 °C.

Attention has been given to the possibility of using waste heat from geothermal power stations for domestic heating.

(d) NEW ZEALAND

Considerable use of geothermal energy for domestic heating has been reported in the town of Rotorua (20,000 inhabitants), a tourist centre with balneo-therapeutic institutions (Kerr et al., 1964; Burrows, 1970; Cooke, 1970).

The town is situated on a high-temperature area and a very large number (about 1,000) of boreholes issuing a mixture of water and steam have been drilled within the city limits. The individual boreholes are connected to single houses, groups of houses or to public building complexes, hospitals, etc., but an integrated district heating system for the whole town has not been organized. This leads apparently to rather wasteful use of the energy, and creates certain disposal problems for the excess hot water. The use of geothermal heating in Rotorua was started about 30 years ago.

A remarkable development is the recent construction of a 100-room tourist hotel in Rotorua that is not only heated with geothermal energy, but also cooled during the hot season by the use of a lithium bromide absorption unit powered with geothermal heat. This installation was commissioned in 1968 (Reynolds, 1970).

(e) U.S.S.R.

Some of the extensive hydrothermal areas of the U.S.S.R. seem to have been discovered and known for very many years as a result of drilling for oil (Tikhonov et al., 1970; Sukharov et al., 1970).

One such deep borehole in Makhach-Kala producing about 23 l/sec of 63 °C water has been used for 22 years for supplying dozens of dwelling houses and industrial buildings with hot water. Today several boreholes are in use supplying heat and hot water to certain districts of the town. One heat distribution station supplies districts with 15,000 inhabitants with 70 l/sec. Geothermal water is also used for hot houses and soil heating in this area.

Systematic prospecting for hyperthermal reserves and drilling for geothermal energy for its own sake did not take place until 1960.

Much attention has evidently been devoted to problems of energy utilization and the development of technically and economically sound systems, quite often involving combined schemes (Kremnjov et al., 1970). A number of experimental systems have been designed and taken into use, some of which will be described later in this paper.

Besides the installations at Makhach-Kala mentioned above, the following have been reported (Lockchine and Dvorov, 1970).

Astarinsk district Azerbaidjan	15 hectares hot houses	46.5 Gcal/h
Zgoudidi town Georgia	Heating of flats, public buildings, industrial buildings, hot houses, swimming pools	50 Gcal/h
Iserbach town Daghestan	District heating for 7,500 inhabitants and industrial uses	6 Gcal/h
Caspillsk town Daghestan	District heating for 5,000 inhabitants and hot water supply	5.0 Gcal/h
Massalinski district	15 hectares hot houses	46.5 Gcal/h
Mendji Georgia	Heating of meteorologic station and agricultural uses	2.0 Gcal/h
Paratounka Kamchatka	Heating of 3 apartment houses with 48 apartments each	0.55 Gcal/h
Ternahir Kamchatka	5 hectares hot houses and soil heating of 5 hectares	19.5 Gcal/h
Zaichi Georgia	Heating of meteorologic station, hot houses and baths	2.1 Gcal/h
Cherkesk Stavropol	District heating for 18,200 inhabitants, industrial uses, hot houses	22 Gcal/h
Total		200.2 Gcal/h

3. Technical aspects

The general technology of house heating, comfort cooling and district heating using conventional sources of energy, such as fossil fuels or electricity, is well known and need not be elaborated here.

Geothermal energy, however, has certain inherent characteristics that have to be taken into account whenever a geothermal system is planned and designed.

1. The thermal fluids have a fixed temperature in each thermal area.
2. Each borehole has as a rule a constant output.
3. The unit cost of the energy produced and/or delivered to the ultimate consumer is predominantly capital costs (depreciation, return on capital, etc., similar to hydro-power).
4. The chemistry of the thermal fluid must be observed.
5. The transportability of the thermal fluids is limited.
6. The unit cost of the energy produced and delivered is dependent on the capacity of the system (scale effect).

Every district heating project must be tailored to the climate of the site. The most significant characteristic of the climate in this respect is the variation of the daily mean outside temperature over the year. Due attention must also be given to the diurnal variation of the outside temperature, and the effect of such factors as wind velocity and solar radiation.

Figure 2 shows qualitatively a typical duration curve for the mean daily temperature over the year.

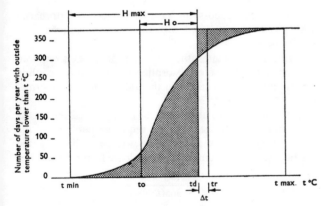

FIG. 2. Typical duration curve for the mean daily temperature over the year.

The objective is to maintain a constant optimum room temperature t_r in the heated buildings, for instance 20-22 °C for apartments. The building receives, in addition to the heat supplied by the system, a certain amount of 'free heat', i.e. heat lost by the occupants, electric lighting and appliances, solar radiation, etc. The free heat increases the room temperature by Δt °C, and accordingly the design room temperature is:

$$t_d = t_r - \Delta t \text{ °C}$$

The power required for heating is directly proportional to the differences between the inside design temperature and the outside temperature:

$$H = k(t_d - t) \text{ kcal/h}$$

where t is the outside temperature (°C).

The shaded area in Figure 2 for $t < t_d$ represents the annual number of degreedays G_h requiring heating, and for $t > t_d$ the degreedays G_c requiring cooling in order to maintain the desired room temperature.

The annual energy demand for heating (or cooling) is directly proportional to the number of degreedays.

The unit energy cost is, as stated before, predominantly capital costs. A geothermal district heating system that has sufficient power to maintain the desired room temperature on the coldest day of the year would have a very low annual load factor, and consequently high unit energy costs. The annual load factor for heating can be defined as:

$$\eta_{min} = \frac{G_h}{365(t_d - t_{min})}$$

where G_h = degreedays for heating above t_{min},
  t_d = the design room temperature °C,
  t_{min} = the mean temperature of the coldest day of the year.

If the geothermal system were designed to maintain the desired temperature to a minimum outside temperature of t_0 only, where $t_0 > t_{min}$, the annual load factor would be:

$$\eta_0 = \frac{G_h'}{365(t_d - t_0)}$$

where G_h' = degreedays for heating above t_0,
  t_0 = the so-called base temperature, °C, i.e. the mean temperature of the coldest day at which the geothermal system can still maintain the desired room temperature.

Reference to Figure 2 shows that this would greatly improve the annual load factor, and reduce the unit cost of the geothermal system correspondingly. The power of the geothermal system would be reduced in the proportion $H_0/H_{max'}$ (Fig. 2), and in the first approximation the unit geothermal energy costs would be reduced in the proportion:

$$\frac{e_0}{e_{max}} = \frac{\eta_{min}}{\eta_0} F \frac{(H_0)}{(H_{max})}$$

where e_0 = unit energy cost for geothermal system with base temperature t_0,
  e_{max} = unit energy cost for geothermal system with base temperature t_{min},
  $F \frac{(H_0)}{(H_{max})}$ is a function that defines the variation of the capital investment for the system in relation to the ratio of installed power (H_0/H_{max}).

The remaining problem is how to handle the heating demand of the relatively few days of the year that have outside temperature $t < t_0$, and days with high wind velocity at which the system may only be able to handle the load at $t > t_0$.

This problem is aggravated by the fact that the water of the geothermal system has constant temperature, and accordingly the efficiency of heat utilization in radiators decreases with increased load.

The simplest solution is to expect the consumers to accept a lowering of the room temperature on the relatively few days of the year with $t < t_0$. This can be tolerated if the cold spells are of short duration and the houses have some heat storage capacity (masonry or concrete construction), but it is not popular with the consumers.

Thermal aquifers, where the temperature allows installation of deep well pumps in the boreholes (150-180 °C or less) can sometimes yield increased production for a limited time by pumping at a considerable draw-down of the water level. This method has been used in Reykjavik, Iceland, for several years (Zoëga et al., 1970).

Heat pumps utilising the heat of the waste water leaving the house systems are being used in Paratounka, Kamchatka, U.S.S.R., to supply additional heat (Lokchine et al., 1970).

It has been suggested (Bodvarsson et al., 1964) to store surplus hot water by injecting it into underground permeable formations if such are available in the vicinity of the system.

The most practical solution is usually to install facilities for peak heating of the water by fossil fuels or electricity as required, perhaps in combination with one or more of the above mentioned methods. This has been done in Iceland, Japan and U.S.S.R. The sum of the energy requirements of the coldest days is so small in comparison with the annual energy needs that the use of expensive peak heating is fully justified.

The selection of the base temperature t_0 for the geothermal system must be taken as a compromise in each case, with due consideration to the climatic premises and the available means of dealing with the peak load.

The demand for heat is not only subject to seasonal variation over the year as discussed above, but also to diurnal variation. Figure 3 shows the variation of the load over 24 hours on two consecutive days in one of the sub-stations of the Reykjavik District Heating System. This comprises the demand for heating and for hot tap water. The general shape of the curve (b) is typical for most days of the year; however, it is of interest to note the difference between the demand of the two days, which shows the influence of 'free heat' due to a few hours of sunshine (a).

In as much as the unit cost of energy in geothermal district heating systems is primarily capital cost, attention should be given to all available possibilities of either reducing the investment or increasing the annual load factor.

At the present state of the art, geothermal power stations operating in geothermal fields with two phase flow have a very low factor of energy utilization, which means that they supply enormous amounts of waste energy. Wherever possibilities permit this should be used for district heating, agricultural or industrial purposes. Combined schemes allowing the use of excess water for heating hot houses and especially for soil heating are practical measures for improving the annual load factor.

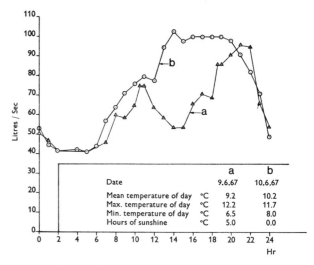

FIG. 3. Variation of the heat load over 24 hours on two consecutive days in one of the sub-stations of the Reykjavik District Heating System.

The above discussions have been concerned primarily with problems associated with the heating of homes and other buildings.

In certain climates, the cooling of premises is required for comfort. The use of absorption equipment for cooling opens up new perspectives for the utilization of geothermal energy (Reynolds, 1970).

The shaded area above for $t > t_d$ in Figure 2 represents the number of degree days G_c requiring cooling. If the energy requirements for cooling were supplied by the district heating system, its annual load factor would be:

$$\eta_0 = \frac{G'_h + xG_c}{365(t_d - t_0)}$$

where G'_h is the number of degree days for heating above base temperature t_0,

G_c is the number of degree days for cooling above the design temperature t_d,

x is the ratio of the energy requirements for each degree day of cooling and heating respectively.

The load factor is thus obviously improved.

Figure 4 shows in a similar way to Figure 2 the main characteristics of three types of climate. Figure 4a shows a subarctic insular climate like that of Iceland, Figure 4b a cold temperate continental climate, and Figure 4c a tropical climate. The last two figures are qualitative but they indicate, as stated above, that there should be an oppor-

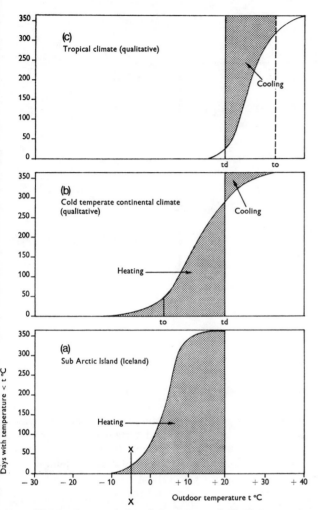

FIG. 4. Main characteristics of three types of climate.

tunity of improving the annual load factor in continental climates by the introduction of comfort cooling, and they raise the question whether an economic basis can be found for district comfort cooling systems in the tropical parts of the world. The tropical countries have an urgent need for cooling in connexion with food processing and storage, and other industrial uses for cooling and heating might be found and developed, comparable with the use of geothermal energy for soil heating and hot houses in colder climates. Combined comfort cooling and industrial schemes might therefore be feasible in the tropics.

4. Description of some geothermal district heating systems

(a) REYKJAVIK DISTRICT HEATING SYSTEM

This is the oldest and still the largest geothermal district heating system in the world.

The principal physical data of the system are summarized in the following tabulation, which is based on published data (Zoëga and Kristinsson, 1970).

1. *Climatic data*

1.1 Mean temperature of the year	+ 5	°C
1.2 Mean temperature of the warmest month (July)	+ 11.2	°C
1.3 Mean temperature of the coldest month (Jan.)	— 0.4	°C

2. *Available heat resources (Dec. 1969)*

2.1 Reykir geothermal area 1,000 m³/h at 80 °C	40 Gcal/h
2.2 Reykjavik geothermal area 1,700 m³/h at 119 °C (average)	135 Gcal/h
2.3 Own peak power boiler plants (oil fired)	30 Gcal/h
2.4 National Power Co. peak power boiler plant (available at electrical off-peak hours only)	20 Gcal/h
Total	225 Gcal/h

3. *Heat load*

3.1 Volume of houses connected	10.3×10^6 m³
3.2 Number of houses connected	8,700
3.3 Heat load at — 10 °C outside and + 20 °C inside	190 Gcal/h
3.4 Specific load at — 10 °C outside and + 20 °C inside	19 kcal/h m³

4. *System data*

4.1 Installed horsepower in pumping plants	5,115 hp.
4.2 Area served by distribution system	11.2 km²
4.3 Length of pipe lines	
4.3.1 Collecting mains	14.2 km
4.3.2 Supply mains	29.1 km
4.3.3 Street mains	125.2 km
4.3.4 House connections	120.2 km
4.4 Average density of population	643 inhabitants/km²
4.5 Average load density	17 Gcal/h km²

5. *Yearly heat production*

5.1 Geothermal energy (1968)	960 Tcal/year
5.2 Peak power stations (1968)	80 Tcal/year
Total	1,040 Tcal/year

The development of the geothermal areas feeding the system has been described earlier in this paper.

Figure 5 shows the present system in principle. The water from the boreholes in the Reykir fields (86° C, 280 l/sec) flows by gravity into collecting tanks (1) from which it is pumped through a 15.3 km long pipeline (two 350 mm diameter pipes) to storage tanks (4), capacity 8,000 m³, on a hill within the city. The pumps are governed by air operated valves on the discharge side operated by level control in the collecting tanks. An oil-fired peak heating plant (2) can raise the temperature of the water on cold days. The water temperature in the main storage tanks is maintained at about 90 °C.

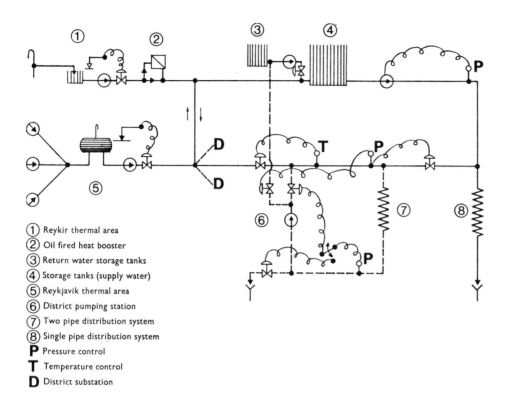

① Reykir thermal area
② Oil fired heat booster
③ Return water storage tanks
④ Storage tanks (supply water)
⑤ Reykjavik thermal area
⑥ District pumping station
⑦ Two pipe distribution system
⑧ Single pipe distribution system
P Pressure control
T Temperature control
D District substation

FIG. 5. Schematic diagram of system.

The boreholes of the Reykjavik fields (5) (temperature range 105-140 °C, average 119 °C, total flow 470 l/sec), are equipped with deep well pumps set at 110-120 m depth. The water in the collecting mains is thus under sufficiently high pressure to prevent flashing in the pipes to the nearest main pumping station. In the pumping stations the water passes through deaerators where controlled flashing occurs in order to remove gases (primarily nitrogen) from the water. The pumps are governed by air actuated valves controlled by the water level in the deaerators.

The temperature of the water supplied to the houses (for heating or hot tap water) is maintained at about 80 °C. In the original system based on the Reykir fields, the water was pumped from the main storage tanks (4) by pumps regulated by the pressure at a suitable point on the system through a single pipe system (8) and the water was wasted to the sewer, after passing through the house system. This system is still in use.

With the advent of the water from the Reykjavik field (temperature 100 °C) it was necessary to adopt two pipe systems (7) whereby a sufficient quantity of return water from the houses was collected in order to cool the high-temperature water (by mixing) to the desired distribution temperature. A number of sub-stations have been built which are fed with high-temperature water and serve combinations of single- and two-pipe systems (6), (7) and (8).

All pumping stations except the deep well pumps are fully automatic. An electronic controlling and monitoring system has recently been commissioned (1969) for remote supervision, data logging and certain control operations (starting and stopping of pumps) from a central control room, for all pumping stations and borehole pumps.

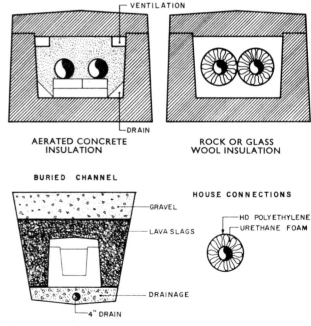

FIG. 6. Street main channels.

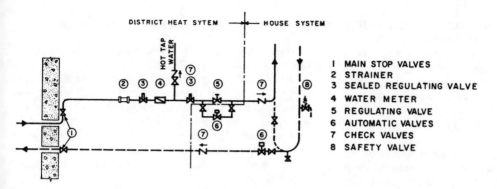

FIG. 7. House connection.

All the system pipe work is underground. It consists of welded black mild steel pipe, laid in concrete channels if the pipe diameter is 75 mm or larger and insulated with rock wool or aerated concrete. Street mains of smaller diameter as well as house connections are black steel pipes insulated with polyurethane foam in the annular space between the steel pipe and an outside protective jacket of high density polyethylene pipe. Figure 6 shows typical cross-sections of the pipes.

Figure 7 shows the standard house connections for a two-pipe system. It should be noted that the geothermal water is used directly as hot tap water. The house connection includes a sealed maximum flow regulator, (3), and an integrating water meter (4) measuring the consumption. The solenoid supply valve (6) is controlled by a room thermostat, and a high limit temperature switch controls the solenoid valve (6) in the return pipe from the radiators.

Figure 8 shows the diurnal load variation in one of

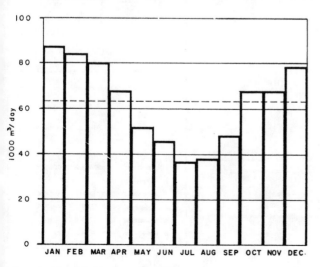

FIG. 8. Monthly water production, 1968.

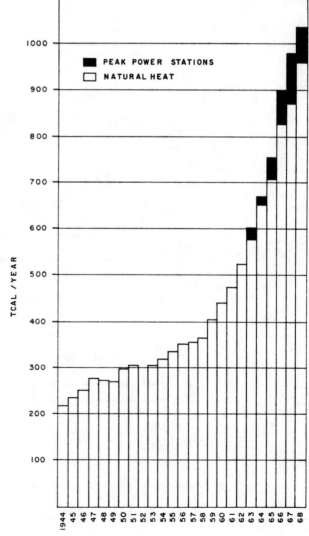

FIG. 9. Yearly heat production.

131

the sub-stations. The daily peak load is generally 15-30% higher than the mean demand over the 24 hours of the day. The monthly variation of the water production is shown on Figure 8, and follows substantially the variation of the monthly mean outside temperature. The annual load factor for the whole system is quite high, corresponding to 5,000 h/yr of the full power of the system, or 57%. For the geothermal system alone the load factor corresponds to 5,800 h/yr, or 66%.

Daily load variation is generally taken care of by the storage tanks, but extreme peaks resulting from cold spells and/or high wind velocities are handled by the storage tanks in combination with peak heating by oil, and intensified pumping from Reykjavik field by the deep well pumps.

Figure 9 shows the growth of the annual heat production 1944-69, and also the portion of energy supplied by the peak heating boilers.

(b) OTHER GEOTHERMAL DISTRICT HEATING SYSTEMS

Lokchine and Dvorov (1970) describe a number of district heating systems in the U.S.S.R. ranging in power from 0.55-50 Gcal/h, using geothermal water in the temperature range 50-90 °C. The thermal waters have different degrees of mineralization. They may therefore sometimes be too aggressive to be used directly for the radiators, in which case heat exchangers are used. Apparently the hot tap water is in most cases heated indirectly by heat exchangers.

Peak heating with fossil fuel fired boilers or electricity is widely used.

Frequently hot houses and/or soil heating is combined with the district heating systems, as well as other industrial or agricultural uses of heat. In this way the annual load factor can be significantly improved especially with seasonal soil heating, which falls outside the periods of maximum demand on house heating (spring and autumn).

Considerable efforts have been made in order to

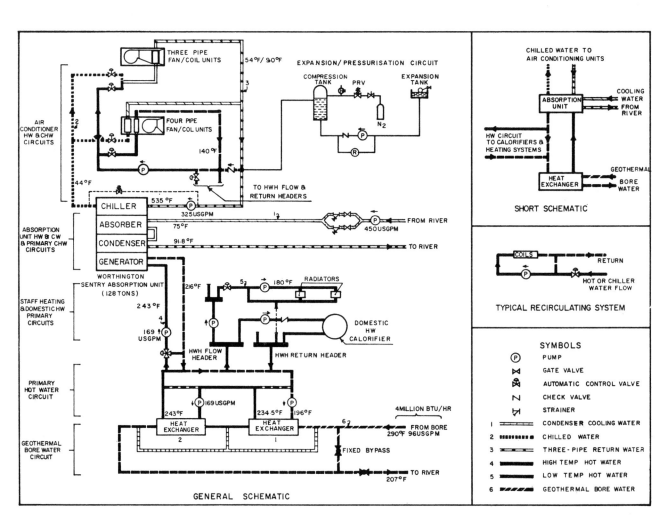

FIG. 10. Scheme of the geothermal heating and air conditioning installation in the Rotorua International Hotel, New Zealand.

increase the efficiency of energy utilization by lowering the temperature of the waste water, for instance by using two-stage heat exchangers for hot tap water, and by using two-stage heating of air.

The most interesting system is perhaps that of Paratounka, Kamchatka, (Lokchine and Dvorov, 1970). The scheme comprises three apartment houses with 48 apartments each (load 0.55 Gcal/h). Geothermal water of 80 °C is used for heating hot tap water and for heating the apartments.

The tap water is heated from + 5 °C in a heat exchanger, cooling the geothermal water to 10 °C.

The apartment houses are heated either by using a central heating system with radiators or by pipes embedded in the concrete of the floors and ceilings (radiant heating). The heating systems are designed for a temperature drop from 80 to 40 °C.

By use of a heat pump, part of the return water of 40 °C can be reheated to 60 °C by extracting heat from the remainder of the return water, which is in turn cooled to 10 °C before being wasted. The 60 °C water is mixed with the geothermal supply water of 80 °C, and the temperature of the mixture can be increased by the use of an electrical peak heating unit.

Use of the heat pump during periods of peak load counteracts the well known fact that the efficiency of heat utilization in geothermal systems using radiators decreases with increased load.

Mashiko and Hirano (1970) report on the use in Japan of various types of pipe laminated with synthetic materials instead of steel pipe in order to avoid corrosion problems associated with mineralized geothermal water.

The use of geothermal energy for cooling for industrial purposes by the use of lithium absorption equipment has been reported in the U.S.S.R. by Tikhonov and Dvorov (1970) who also point out the possibilities of using the equipment for cooling in summer and heating in winter (heat pump application) in the southern regions of the U.S.S.R. Mashiko and Hirano (1970) also mention industrial refrigeration with geothermal energy in Japan.

However, the most interesting installation in the present context is the geothermal heating and air conditioning installation in the Rotorua International Hotel, New Zealand, reported by Reynolds (1970), see Figure 10.

The system is designed for the extreme climatic temperatures of − 4 °C and + 30 °C (25 °F and 85 °F). The maximum heating load is 0.5 Gcal/h (2,000,000 BTU). Two calorifiers for the heating of tap water have a combined maximum demand of 0.5 Gcal/h, but the house heating (or cooling) has preference in the case of coincident peak demand on both systems. A 130 ton (0.39 Gcal/h) lithium bromide absorption unit supplies the cooling for the air conditioning and requires a heat input of 0.575 Gcal/h. The specific energy requirement of the absorption unit is therefore 1.47 kcal heat per 1 kcal of cooling.

The heat energy is supplied by a borehole producing at temperatures above 150 °C and a pressure of about 6 atg. The heat is transferred by heat exchanger to fresh water in closed circuits which is heated to 120 °C, and supplies heat to the radiators, tap water heaters and the absorption unit.

5. Some economic aspects

The economic feasibility of using geothermal energy for heating (or cooling) depends on whether it can compete with other available sources of energy such as fossil fuels, electricity, etc.

Such comparison should be based on the total cost to the ultimate user per unit of net energy utilized. This means comparing costs of entire systems, taking into account the desired return on invested capital of all plant used, direct operating costs, annual load factors, efficiency of energy utilization, etc.

The principal factors that affect the economic feasibility of geothermal district heating systems are the following:
1. The drilling costs per unit energy production ($/Gcal/h).
2. The temperature of the available geothermal fluids.
3. The distance from the geothermal field to the centre of gravity of the market.
4. The load density of the market (Gcal/h/km²).
5. The annual load factor of the system.
6. The power of the system (Gcal/h).

The drilling costs per unit of energy production govern the cost of the energy ex borehole. The temperature of the fluid, the distance of transmission to the market and the load density are the main factors that influence the transport and distribution costs. The influence of the annual load factor has been discussed earlier in this paper as well as the scale effect which is related to the power of the system.

The specific drilling costs ($/Gcal/h) can differ within wide ranges. They were thus about 2,100 U.S. $/Gcal/h in one high temperature field in Iceland (Southern Hengill) and 16,700 U.S. $/Gcal/h in the Reykjavik geothermal field (low temperature area 120 °C average) (Bodvarsson et al., 1964). Kremnjov et al. (1970) show the variation of drilling cost with depth under various geological conditions in the U.S.S.R.

The economic transportability of geothermal fluids is relative low and highly dependent on the temperature of the fluid. Several installations are in use where water of less than 100 °C is transported by pipeline over 10-20 km. Water of 150-180 °C can probably be transmitted for house heating purposes over 50-75 km, provided a large concentration market is available (more than 200 Gcal/h).

Installations are in use where the load density of the market is in the range of 10-17 Gcal/h/km².

The specific capital investment ($/Gcal/h) for geothermal district heating systems varies within wide ranges depending on local conditions. Zoëga *et al.*, (1970) report the average costs for the Reykjavik system, based on present day methods and equipment as follows:

Heat production	29,400 U.S. $/Gcal/h
Distribution system	58,000 ,,
Total	87,400 U.S. $/Gcal/h

and estimate that the replacement value of the present system (225 Gcal/h) is of the order of U.S. $17 million.

The energy price paid by the customers is 0.16 $/m^3 of water at 80 °C. Based on average utilization, this corresponds to 3.80 $/Gcal and is broken down as follows (Palmason *et al.*, 1970):

Drilling	0.73 $/Gcal
Main pipelines	0.42 ,,
Storage	0.15 ,,
Distribution	2.50 ,,
Total	3.80 $/Gcal

The savings in oil by the use of geothermal energy depend on the annual load factors. The following figures are reported:

Cherkest (U.S.S.R.)	680 tons oil/Gcal year
Reykjavik (Iceland)	870 ,,
Paratounka (U.S.S.R.)	1,090 ,,
Caspillok (U.S.S.R.)	1,440 ,,

The Reykjavik district heating system thus saves about 150,000 tons of oil annually that would have had to be imported, and the annual cost of heating for the customers is only 60% of the cost of heating with oil.

Acknowledgement

The author is indebted to his partners in Vermir H/F, Consulting Engineers, Iceland, for free use of data from their files in preparing this paper.

Bibliography

BODVARSSON, G. 1954. Laugarhitun og rafhitun (Geothermal and Electrical Heating). *Timarit Verkfraed. Islands*, vol. 39 no. 1. (In Icelandic.)

——; ZOËGA, J. 1961. Production and Distribution of Natural Heat for Domestic and Industrial Heating in Iceland. *Proc. U.N. Conf. on New Sources of Energy, Rome 1961*, vol. 3 (Paper G/37). New York, United Nations.

——; PALMASON, G. 1961. Exploration of Subsurface Temperature in Iceland. *Proc. U.N. Conf. on New Sources of Energy, Rome 1961*, vol. 2 (Paper G/24). New York, United Nations.

——; EINARSSON, S. S. 1964. Jardhiti til húshitunar og idnadar (Geothermal Energy for Domestic and Industrial Heating), *Rádstefna isl. verkfraedinga, Reykjavik 1964* (Proc. Conf. of Icelandic Engineers, Reykjavik, 1964). Reykjavik, Verkfraedingafélag Islands. (In Icelandic.)

BOLDIZSAR, T. 1970. Geothermal Energy Production from Porous Sediments in Hungary. *U.N. Symp. on Geothermal Energy. Pisa 1970.*[1]

BURROWS, W. 1970. Geothermal Energy Resources for Heating and Associated Applications in Rotorua and Surrounding Areas. *U.N. Symp. on Geothermal Energy. Pisa 1970.*[1]

COOKE, W. L. 1970. Some Methods of Dealing with Low Enthalpy Water in the Rotorua Area of New Zealand. *U.N. Symp. on Geothermal Energy, Pisa, 1970.*[1]

EINARSSON, S. S. 1970. Utilization of low enthalpy water for space heating, industrial, agricultural and other uses. Rapporteur's report, *U.N. Symp. on Geothermal Energy, Pisa 1970.*[1]

KERR, R. N.; BANGMA, R.; COOKE, W. L.; FURNESS, F. G.; VAMOS, G. 1961. Recent Development in New Zealand in the Utilization of Geothermal Energy for Heating Purposes. *Proc. U.N. Conf. on New Sources of Energy, Rome 1961*, vol. 3 (Paper G/52). New York, United Nations.

KOMAGATA, S.; IGA, H.; NAKAMURA, H.; NINCHARA, Y. 1970. The Status of Geothermal Utilization in Japan. *U.N. Symp. on Geothermal Energy, Pisa 1970.*[1]

KREMNJOV, O. A.; ZHURAVLENKO, V. Ja.; SHURTSHKOV, A. V. 1970. Technical-Economical Estimation of Geothermal Sources. *U.N. Symp. on Geothermal Energy, Pisa 1970.*[1]

LOKCHINE, B. A.; DVOROV, F. M. 1970. L'élaboration expérimentale et industrielle de l'alimentation en chaleur et en énergie géothermale en U.R.S.S. *U.N. Symp. on Geothermal Energy, Pisa 1970.*[1]

MASHIKO, Y.; HIRANO, Y. 1970. New Supply Systems of Thermal Springs to Wide Areas in Japan. *U.N. Symp. on Geothermal Energy, Pisa 1970.*[1]

PALMASON, G.; ZOËGA, J. 1970. Geothermal Energy Development in Iceland 1960-1969. *U.N. Symp. on Geothermal Energy, Pisa 1970.*[1]

RAGNARS, K.; SAEMUNDSSON, K.; BENEDIKTSSON, S.; EINARSSON, S. S. 1970. Development of the Námafjall Area Northern Iceland. *U.N. Symp. on Geothermal Energy, Pisa 1970.*[1]

REYNOLDS, G. 1970. Cooling with Geothermal Heat. *U.N. Symp. on Geothermal Energy, Pisa 1970.*[1]

SIGURDSSON, H. 1961. Reykjavik Municipal District Heating Service and Utilization of Geothermal Energy for Domestic Heating. *Proc. U.N. Conf. on New Sources of Energy, Rome 1961*, vol. 3 (Paper G/45). New York, United Nations.

SUKHAREV, G. M.; VLASOVA, S. P.; TARANUKHA, T. K. 1970. The utilization of Thermal Waters of the Developed Oil Deposits of the Caucasus. *U.N. Symp. on Geothermal Energy, Pisa 1970.*[1]

TIKHONOV, A. N.; DVOROV, I. M. 1970. Development of Research and Utilization of Geothermal Resources in the U.S.S.R. *U.N. Symp. on Geothermal Energy, Pisa 1970.*[1]

ZOËGA, J.; KRISTINSSON, G. 1970. District Heating for Old and New Town Development. Paper presented at the First International District Heating Convention, London 1970.

1. Published by the Istituto Internazionale per le Ricerche Geotermiche, Lungarno Pacinotti 55, Pisa, Italy.

Industrial and other applications of geothermal energy

(except power production and district heating)

B. Líndal

Consulting Engineer
(Iceland)

Introduction

Apart from power production and district heating, geothermal energy has two further major fields of usefulness, namely industry and agriculture. There are also the balneological and recreational fields. It is found that industrial applications largely require the use of steam, while agricultural users may equally well use geothermal water in most cases. Although there are already important applications in these fields, future developments are expected to be still more impressive, and it is likely that the real potentialities of natural heat have hardly been touched upon as yet. There is little doubt that the pioneering enterprises in these fields will prepare the way for a spate of new developments.

The harnessing and application of geothermal energy for industrial and agricultural purposes is a new activity which at present is the subject of a great deal of research and process development. Till now, the greatest emphasis has been placed on exploration but less progress has been made on the technical aspects of applying the geothermal fluids. The characteristics of such fluids are such that a special effort is needed to make them widely applicable.

Broadly speaking, there is a resemblance between the characteristics of this new source of energy and hydro-electric power which caused the upsurge of large electric-power-consuming industries at the turn of the century. At that time, low-cost hydropower triggered a major development in the power-hungry metallurgical industries in places like Norway and Canada. Just as hydropower was very inexpensive in some places, major quantities of natural steam may be available at a fraction of the cost of steam from conventional sources. And just as electrolytic processes could absorb tremendous amounts of hydropower, there are heat-consuming processes which could absorb vast quantities of cheap calories. However, geothermal fluids cannot be transported over long distances, and in this respect they are of much more limited application than electric power. New activities must therefore be developed where geothermal energy is available. Hence, this new source of energy can provide a local stimulus to activities which are fortunate enough to be within its range, or it can create new activities which would otherwise be out of the question.

Industrial applications

Geothermal energy may be used in a number of ways in the industrial field. It may be simple process heating, it may be drying or distillation in every conceivable fashion, it may be refrigeration or it may be de-icing or tempering in various mining and material handling operations. Geothermal fluids may themselves also furnish useful raw materials in some cases. Some thermal waters contain salts and other valuable chemicals, while the steam may contain some industrially useful non-condensable gases (although their usefulness has been of minor importance so far).

In this section, applications involving only energy supply will first be discussed, followed by mixed processes and the recovery of raw materials from geothermal fluids.

The amount of steam which may be applied per unit weight of product in a number of industrial processes is shown in Table 1. Most of the values for steam consumption in the first column are quoted from Chilton (1960) and represent common industrial practice. It should be noted that the values for steam consumption are based on the use of fossil fuels, and may not give correct indications of the possible use of geothermal steam.

But the conventional steam consumption per unit weight of product does not give a satisfactory measure of its importance in the production process. A much more reliable indicator is the amount of steam used per unit *value* of the product shown in the third column. For reference, the product values are recorded in the second column of the table in terms of U.S. cents per pound according to recent listings, where available. For better understanding of the meaning of this ratio, compare for instance the recorded values for ascorbic acid (vitamin C) with those for fresh water.

TABLE 1. The specific consumption of steam and the steam used per dollar value in some established processes

Product and process	Steam requirements	Product value	Steam per unit product value
	lb steam/lb	cents/lb	lb steam/ $value
Heavy water by hydrogen sulphide process	10,000	3,000	333
Ascorbic acid	250	250	100
Viscose rayon	70[1]	75	93
Lactose	40	14	286
Acetic acid from wood via Suida process	35	10	350
Ethyl alcohol from sulphite liquor	22	7	314
Ethyl alcohol from wood waste	19	7	271
Ethylene glycol via chlorohydrin	13	13	100
Casein	13	56	23
Ethylene oxide	11	15	73
Basic Mg carbonate	9	11	82
35% hydrogen peroxide	9	18	50
85% hydrogen peroxide from 35% H_2O_2	$4\frac{3}{4}$	—	—
Solid caustic soda via diaphragm cells	8	3	266
Acetic acid from wood via solvent extraction	$7\frac{1}{2}$	10	75
Alumina via Bayers process	7^2	3	234
Ethyl alcohol from molasses	7	7	100
Beet sugar	$5\frac{3}{4}$	10	58
Sodium chlorate	$5\frac{1}{2}$	9	61
Kraft pulp	4 1/5	6	70
Dissolving pulp	4 1/5	—	—
Sulphite pulp	$3\frac{1}{2}$	6	58
Aluminium sulphate	$3\frac{1}{2}$	2	175
Synthetic ethyl alcohol	3	7	43
Calcium hypochloride, high test	$3\frac{1}{3}$	3	111
Acetic acid from wood via Othmer process	$2\frac{3}{4}$	10	28
Ammonium chloride	$2\frac{3}{4}$	6	46
Boric acid	$2\frac{1}{4}$	5	45
Soda ash via Solvay process	2	$1\frac{1}{2}$	133
Cotton seed oil	2	10	20
Natural sodium sulphate	1 4/5	$1\frac{1}{2}$	120
Cane sugar refining	1 2/3	10	17
Ammonium nitrate	$1\frac{1}{2}$	$3\frac{1}{2}$	43
Ammonium sulphate	1/6	$1\frac{1}{2}$	11
Fresh water from sea water by distillation	1/12	1/60	500

1. Shreve (1956) quotes 150 lb steam per pound.
2. Has declined in recent years in most cases.

Armstead (1969) has recorded the cost of steam produced by fossil fuels. His conclusion is that in power plants ranging from 1 to 500 MW, the cost of steam at the point of use would range from 38 to 103 U.S. cents per million BTU. Since a long ton of steam represents approximately $2\frac{1}{2}$ million BTU, this gives an average figure of about $1.75 per ton of fuel-raised steam as a useful preliminary yardstick, although big consumers may have lower costs and small ones higher costs.

In the power-consuming industries, the cost of electrical power becomes a major consideration in plant location when it accounts for more than about 10% of the value of the manufactured goods. In such cases a thorough consideration must be given to specially cheap sources of power, and a number of disadvantages may have to be balanced against gains in this respect. If similar considerations apply to steam-consuming industries, it will be seen from Table 1 that several products would be in a similar position. If the yardstick for steam cost is taken at $1.75 per ton of steam consumed, then 128 pounds of steam per one dollar value would account for 10% of the value of manufactured goods. In Table 1, 10 items out of a total of 32 exceed this figure. But even 4 to 5% of the value of goods for steam costs can be of serious consequence. In the range 4 to 10% 13 more items of Table 1 would be included. Yet, it is not likely that any possible economies in steam-consuming processes based on present practices in the established industries will yield more than a fraction of the possibilities of geothermal heat.

Consideration will now be given to some examples of current practice in this field, as well as further applications which are being investigated and developed.

PULP AND PAPER; WOOD PROCESSING

There are two principal methods for processing essentially pure cellulose pulps out of wood, namely the Kraft or sulphate pulp process and the sulphite process. According to Shreve (1956) the Kraft process is responsible for the major part of the pulp manufactured at the present time. Almost any kind of wood may be used, although coniferous woods are mostly employed. The process is essentially chipping of the wood, digesting the chips by steam in the presence of chemicals, separation of the black liquor from the cellulose and, finally, recovery of the required grade of pulp by washing, drying, etc. The black liquor is usually evaporated and the end fed to a furnace where the inorganic chemicals may be recovered as a residue and re-cycled to the process.

In the sulphite process, the chips are also digested by steam in the presence of chemicals, but the waste liquor, called sulphite liquor, is rarely recovered. This waste contains more than half of the raw material entering the process, largely as dissolved organic materials. For the sulphite process, spruce is the wood most commonly employed although appreciable amounts of hemlock, balsa and others are also used.

The pulp and paper mills of the Tasman Company in New Zealand were the first major industrial development to use natural steam, and the site of the mills was selected with that in mind. For some years now, this activity has been using almost 400,000 lb of natural steam per hour. The following description is obtained from Smith (1970).

The mills of the Tasman Pulp and Paper Company in Kawerau produce newsprint, pulp (using the Kraft process) and sawn timber. The mills have been erected close to the source of geothermal energy at the Tarawera River. Figure 1 shows the layout of the boreholes, the steam lines and the plant.

The bores produce wet steam, which is at present flashed before transmission. The steam is transmitted by two principal pipelines; one of 12-in carrying up to 80,000 lb/h of steam at 200 psig and one of 24-in, having a capacity of 320,000 lb/h of steam at 100 psig.

At the mills some of the natural steam is used directly, and some indirectly to produce clean steam. The high-pressure steam, 200 psig, is used directly for timber drying and for the log-handling equipment, while it is finally fed to the heating coils of a steam generator having a capacity of 55,000 lb/h of 150 psig clean steam, presumably largely used for the pulp digesters.

Up to 320,000 lb/h of 100 psig steam is used directly for several purposes, while some is used for another steam generator having a capacity of 45,000 lb/h clean steam at 50 psig. The surplus is fed to a 10 MW non-condensing turbo alternator exhausting to atmosphere. Up to 100,000 lb/h of this exhaust steam is then used in the black liquor pre-evaporator. The 10 MW power unit is generally only partly loaded, but since it operates in parallel with other units fed with boiler steam, it can take over the load in case of failure. At full capacity it requires 319,000 lb/h of steam.

Plants of this type have to recover inorganic chemicals from the black liquor for economical reasons. The first stage is usually multiple effect evaporation for which natural steam is used. After that there is a further dewatering and the organic materials are finally burned off in a special furnace and thus separated from the inorganic materials which are left for recovery. Since a considerable part of the organic matter in the wood is dissolved in the black liquor this represents a major source of energy when the heat of combustion is recovered, as is customary. The Kawerau plant produces thus 350,000 lb/h of high-pressure steam from the black liquor and wood waste. Coal and oil are also said to be used, but probably largely for emergencies. This high-pressure steam appears to be the main source of electrical power in the plant.

Since the Kraft process has thus proved successful in using natural steam it might be worth while to seek a location where natural steam could be used for the sulphite process in view of markets, raw materials and other pertinent factors. For one thing the sulphite process makes use of sulphur, which would to some extent be available from the natural steam itself in the form of H_2S; also the natural steam would increase the chance of recovering useful by-products from the waste sulphite liquor.

Most wood pulp manufacturers also produce finished paper from the pulp. This also takes substantial amounts of steam in the general process and drying operation.

Another use of wood pulp is for the manufacture of viscose rayon. One source quotes that 70 lb of steam are used per pound of viscose yarn; another quotes even 150 lb. This would suggest that substantial benefit could be reaped through the use of natural steam if other pertinent factors are also favourable, since steam cost is an important factor in the total manufacturing cost.

Timber seasoning by natural heat appears to be common. Such operations are mentioned by Burrows (1970) and Líndal (1964a). Burrows also mentions veneer fabrication, and a number of other operations with timber, which could benefit from the use of natural heat, or already do so.

SUGAR PROCESSING, BORIC ACID, SALTS FROM SEA-WATER, HEAVY WATER, FRESH WATER AND OTHER PROCESSES

There is a small salt plant in Japan which makes use of geothermal energy for evaporating seawater. As stated by Komogata et al. (1970) there are now plans for a new plant

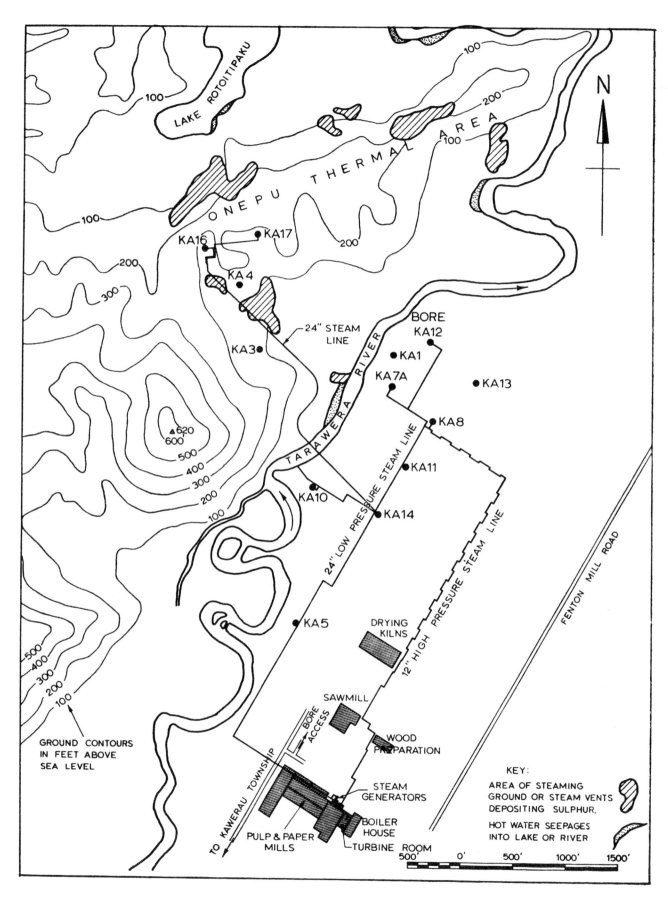

FIG. 1. The pulp and paper mills at Kawerau, New Zealand (from Smith, 1970).

designed for 100,000 tons of table salt yearly. Electro dialysis combined with evaporation will be used.

Líndal (1970a) describes a new process being developed for the production of magnesium chloride and soda ash from seawater and salt. Magnesium hydroxide is precipitated out of the sea by lime, as in current industrial practice. The magnesium hydroxide is then dissolved by the addition of CO_2 and thereby a magnesium bicarbonate solution is formed. By a conversion, using a cation exchange resin, and by the introduction of NaCl two solutions are produced, one of magnesium chloride and the other of sodium bicarbonate. Solids must be recovered from both of these solutions and much heat energy is needed for this. Natural heat would be very suitable for this purpose, and the process is being developed with that in mind.

In Larderello, Italy, boric acid used to be produced from the natural steam itself until recently. This recovery of boric acid involved considerable steam loss, so that the practice had to be stopped to enable fuller use to be made of the steam for the important electric power production. Nevertheless, boric acid is still produced at Larderello, but from imported ores. This operation also requires steam, and it is believed that about 30 tons per hour of natural steam are still being used there for this purpose.

Hallsson (1971) reports an extensive research and development programme involving the drying of various seaweeds in Iceland by the operation of a prototype conveyor dryer using geothermal water at 100 °C as the source of energy. Plans for commercial operation have been made. The same dryer has been used successfully for the drying of grass.

There are some minor, but quite important, uses of natural heat. Thus the use of geothermal energy for curing light aggregate cement building slabs has been reported in Iceland, washing and drying of wood takes place in two or more countries, and stock fish has been dried in this way. Drying of peat has also been considered.

A very interesting field for the application of geothermal energy may be found in sugar processing. Cane sugar, for instance, goes through two process steps, each requiring considerable amounts of steam; first the production of raw sugar and then its refinement. The heat energy is largely used for evaporation in multiple effect evaporators. With beet sugar still more energy is used. Thus, 6,000 to 10,000 BTU are consumed in the form of coal to produce power and process steam for one pound of sugar, according to Shreve (1956).

Although bagasse is used for fuel in most raw sugar mills, it is yearly becoming a more valuable raw material. Many important products may be obtained from it, so it is becoming too valuable for burning.

Another interesting and closely related field is found with fermentation processes based on molasses. Among the products are ethyl alcohol, butanol, acetone and citric acid, all of which may benefit by the availability of liberal amounts of steam. It is reported that natural heat is already used in Japan for brewing and distillation.

Some industrial activities may involve the reverse action of heat; that is refrigeration. This may be achieved through the use of absorption refrigeration, which in turn requires thermal energy as a driving force. Tikhonov and Dvorov (1970) mention the possibility of using refrigeration by natural heat in the production of synthetic rubber, ammonia, protein and vitamins. Metallurgical plants also make great use of refrigeration. Freeze drying of foodstuffs by natural heat has been studied as reported by Ludviksson (1970). Cooling by the use of liquefied air has been proposed by Einarsson (1970), who also points out that the metallurgical industries use pure oxygen which could be obtained by these means.

The production of heavy water has interested scientists working with geothermal energy for more than two decades, but for various reasons no such plant has been built as yet. Valfells (1970) has given an up-to-date account of the situation in this field. His conclusion is that in spite of several competitive processes, the hydrogen sulphide-water isotope exchange process is still the most economical in large heavy water plants, and that this should be the process used in conjunction with geothermal steam. Where other conditions are also favourable he concludes that geothermal energy has a considerable advantage by comparison with other energy sources.

The possibility of using geothermal energy for the production of fresh water has been raised several times in the past. Bodvarsson mentions it (1964), and Armstead dealt with this subject at the International Conference on Water for Peace, 1967. Wong has also discussed this matter (1970). It is pointed out that the fresh water may either come from an outside source or originate in the geothermal fluid itself. Since the practice of purifying water from an outside source by distillation is already highly developed, natural steam could undoubtedly be used instead of conventional steam. The second alternative will be further discussed later in this article.

MINING AND UPGRADING OF MINERALS

Perhaps one of the most interesting and important future roles of geothermal energy is in the field of mining and upgrading of minerals. The process heating available from geothermal fluids may even be of such importance that mineral deposits which were hitherto worthless may become economically attractive. Such is the case with the recovery of diatomaceous earth in Northern Iceland.

Recently a huge deposit of diatomaceous earth was discovered under water in Lake Myvatn. The mineral proved to be of satisfactory quality for the production of diatomite filter aids, but nobody had been able to recover this material economically by wet mining and hitherto the useful mines were those which could be operated by dry excavation techniques. The difficulty with water entrainment is that this highly porous material tends to retain 3 to 4 times its own weight of water, even after filtration.

FIG. 2. The diatomite plant at Myvatn, Iceland, and the steam supply system.

Fortunately, the Námafjall geothermal area was only 3 km away from the lake. This made possible a commercial diatomite plant which started to operate in late 1967. The steam supply system has been described by Ragnars *et al.* (1970), and the use of the geothermal steam in the plant by Líndal (1970b).

The diatomite plant of Kísilidjan Ltd. was built on a smooth lava field close to the Námafjall geothermal area. The diatomaceous earth is dredged from the bottom of the lake by a suction dredger, and the diatomaceous slurry transmitted by pumping through a 3 km pipeline to the plant site. Up to 50 tons per hour of steam at 10 atg/183 °C may be transmitted from the boreholes some 600 m away. A general layout is shown in Figure 2. The present capacity of this plant is 42,000 tons of dried diatomaceous earth which subsequently is turned into 24,000 tons of diatomite filter aids, the final product. Dredging in the lake is done only in the summer, while the plant runs all through the year.

The usefulness of the natural heat begins at the lake. This place lies just beyond the 65° parallel at an altitude of 277 m above sea level. The average temperature stays below freezing point 6½ months during the year, but freezing may in fact occur at any time. A multitude of warm springs outcrop along the shore, and some in the lake itself. As a result, long stretches of the shore are permanently icefree. This by itself helps to prolong the dredging season. Besides, it is helpful to have warm water in abundance for mixing with the diatomaceous slurry during trans-

mission to the plant. A relatively constant and slightly raised temperature helps in pumping, as well as screening and classifying the slurry which coincides with transmission to the storage reservoirs at the plant.

The reservoirs at the plant contain settled diatomaceous earth covered with water. A suction dredger operates over this water, delivering slurry to a storage tank at the plant all the year around. Normally the water would be frozen around this dredger for the greater part of the year. However, this machine needs freedom of movement and must have a considerable ice-free area in order to function satisfactorily. This problem was solved in the way nature itself does it at the lake. Up to 20 l/sec of warm water from wells at the plant site (30 °C) are heated to about 85 °C by direct injection of natural steam, and discharged in the area immediately around the dredger and floating pipeline during the winter season. Thus, up to 7 tons of steam are used here per hour to make this operation possible all the year around.

Figure 3 illustrates the process used for further treatment. The next application of steam is for preventing the contents of the storage tank from freezing. For this purpose a steam heated coil is placed inside the tank around the edges of the bottom.

The vacuum filters at the plant are fed with a 10-12% slurry at 80-90 °C. Thus, 20 l/sec must be heated by direct injection of natural steam. This takes about 8 tons per hour all the year around.

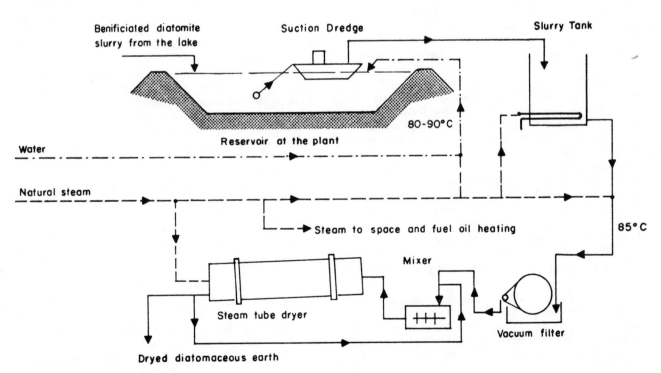

Fig. 3. The utilisation of natural steam in diatomite plant, Iceland.

The last step, for producing a suitable feed for the filter-aid section of the plant, is the drying of the cake obtained from the filters. The dryers are of the rotary steam tube type and natural steam is used directly. About 30 tons per hour of steam are used for this purpose. Finally steam is used for space heating and the heating of the fuel oil used for calcining in the filter aid section of the plant. The total steam consumption in the plant varies according to the season from 40 to 50 tons per hour.

The proposed use of geothermal water for mining operations in the U.S.S.R. is discussed by Tikhonov and Dvorov (1970). They explain that warm water for mining purposes is urgently needed for the exploitation of fossil fuels and for other mining purposes in the regions of permanent frost in the north-eastern part of the U.S.S.R. These regions contain immense resources of gold, tin, wolfram and other minerals, but at present these resources can be exploited only 3 to 5 months in the year owing to the severe climate. Some thermal springs are known in these areas but Tikhonov and Dvorov explain that more water could be obtained by drilling. The thermal water could be used for human comfort and subsistence by farming, as is already practised to some extent, and also for mining.

It is necessary to have water at 20-30 °C for all-year around mining, they explain, but not for seasonal work. By the use of warm or hot water, the pulp to concentration mills could be heated, increasing the effectiveness of ore flotation, and the ore air could be heated and humidified in winter.

Another possible application for natural steam is in the production of alumina from bauxite. This possibility was studied in Iceland some years ago but not put into practice owing to unfavourable raw material and market conditions (Líndal, 1964a). This application could have merit there later on, and the idea could be used elsewhere, where conditions are favourable. The Bayers process is usual for producing alumina, which is the raw material for aluminium smelting. This method involves the use of great quantities of steam; natural steam would be quite suitable. It may be used both for the digestion of the ore in heated reactors, and for the concentration of the resulting solution in multiple effect evaporators. An indication of the importance of the steam requirements for processing can be obtained from Table 1. But this is *only* an indication because the design should be optimized, especially in respect of the natural steam available and the raw material in question. It should be noted also that in recent years considerably less steam based on fossil fuel has been used in some cases than quoted in the table, although it is quite substantial in every case.

Many high temperature geothermal areas have some deposits of elemental sulphur. Although this kind of sulphur used to be mined centuries ago, and still is at least intermittently in some parts of the world, it is rarely important nowadays due to the small and scattered deposits. But one such deposit of considerable size is now being studied in New Zealand according to Smith (1970) and it is hoped that geothermal steam may help by providing process heating in the refining of the ore. Elemental sulphur on a small scale is being recovered directly from the fumes of a volcano in Japan, as reported by Komogata *et al.* (1970).

MATERIALS IN GEOTHERMAL FLUIDS AND THEIR RECOVERY

The principal component of all geothermal fluids is of course water. When ground reservoir temperatures do not exceed about 100 °C, the extent of mineralization may be so low that the water is potable. Warm springs include many of the most famous health mineral springs in the world, where the water may be bottled for distribution. Such operations are spread the world over.

In the case of high enthalpy fluids, the water is not directly potable owing to the presence of contaminants. However, this water contains enough energy for self-evaporation, and it may very well yield the least expensive water conversion.

Wong (1970) describes in detail the development programme which has been initiated for determining the possibilities for producing fresh water, electric power and minerals from the highly mineralized geothermal brine in the Imperial Valley, California. According to the timetable presented, this study will be completed by 1975. It involves the possibilities of producing fresh water for irrigation and other pertinent uses, so as to promote agriculture in one of the world's richest agricultural areas, where crops may be grown throughout the year. The possibility of recovering minerals seems equally impressive, although there are abundant reserves of common salt in this part of the world. Werner (1970) gives a further account of the possibility of recovering minerals. The useful ones might be calcium chloride (already recovered to some extent) potassium chloride and others.

Líndal (1970a) described plans for extracting common salt, calcium chloride, potassium chloride and bromine from geothermal brine in the Reykjanes peninsula, south-western Iceland. The reservoir salinity of this brine is that of seawater, but appreciable concentration occurs by self-evaporation before the brine is discharged from the bore-holes. The steam released is enough for the complete recovery of the solids. Fresh water has little bearing in this project, since plenty of that may be had anyway, but common salt is essential. This project is now in the last stages of preliminary exploration.

Shcherbakow and Dvorov (1970), Tikhonov and Dvorov (1970) and Sukharev *et al.* (1970), discuss the future possibility for extracting chemicals from thermal waters in the U.S.S.R. Some thermal waters in the U.S.S.R. have a high content of bromine and iodine, but the present production satisfies normal requirements. Increased production of these chemicals is mentioned in order to satisfy new objectives. Interest is also shown in alkaline metals, boron, lithium and a number of trace elements.

Very interesting possibilities with minerals were pointed out by Ilukor (1970) in Uganda. Here, many thermal springs contain significant amounts of sodium bicarbonate. The production of common salt, as a preliminary step in utilizing these springs, has been reported.

The self-evaporative quality of high enthalpy thermal water is very significant. Thus, there is little doubt that this will make economic recovery of minerals possible at much lower brine concentration than is ordinarily the case.

Geothermal fluids always contain some volatile components which accompany the steam. These appear as the noncondensable gases and sometimes minor quantities of boron and ammonia are also involved. Only rarely has economic recovery of these components been possible. More often these materials are a nuisance that somewhat reduces the efficiency of the steam as a source of energy. In any case, such recovery has generally not been possible independently of the utilization of the energy. Instead, the recovery tends to be a by-product of some other main production. Only in the very early recovery of boric acid in Italy was this done independently.

CO_2 is the predominant component of the non-condensable gases, H_2S is usually the second, elemental hydrogen the third, and there are also small amounts of methane, nitrogen, ammonia and argon. The amount of these materials in the steam depends on many geological conditions and doubtless also upon the way the drilling is planned in some cases. However, the amount seems to vary between 1 and 10 litres of total non-condensable gas per kg of steam in the great majority of cases, where steam is harnessed for any length of time.

The greatest experience in utilizing these gases was doubtless obtained in Larderello where these gases were recovered for many years from steam used in the production of electric power. This is described and discussed by Mazzoni (1948) and by Garbato (1961) and Lenzi (1964). Besides boric acid, ammonium bicarbonate, ammonium sulphate and sulphur used to be recovered, but by 1970 no such production was mentioned. The only known record of present use of these gases comes from Japan. Here Hayashida and Ezima (1970), report that sulphuric acid is produced from gases at the Otake geothermal power plant. But even if these gases are of small importance compared to the energy, they may very well be of economic value in some localities, especially if the steam is used for industrial purposes. Such activities often require sulphur or sulphuric acid and sometimes carbon dioxide.

Agricultural and related uses

The agricultural use of natural heat is already of major importance in several countries, especially in greenhouse applications. The temperature of the geothermal fluid may be as low as 60 °C for greenhouses and rarely exceeds 100 °C. For soil warming in the open, which is gaining momentum along with greenhouse applications, the temperature needs to be no more than 40 °C, and even lower initial temperatures can be used. Fish farming may make use of fluids of still lower temperatures, since the environmental temperatures for fish are quite moderate.

GREENHOUSES AND SOIL WARMING

In temperate climates the cost of operating greenhouses with fossil fuels usually amounts to 15-20% of the value of the products. Geothermal energy is therefore of great economic interest, and it is not surprising that development in this field has been rapid.

There has been an impressive expansion in greenhouse activity in Hungary according to Boldizsar (1970). By the end of 1969 there were 400,000 m² of greenhouses, and there was a plan for doubling this in 1970.

Facca (1970) remarked that the first large scale development in the use of geothermal energy for space heating and for greenhouses occurred in Iceland. By 1960 there were already 95,000 m² of greenhouse area covered with glass as reported by Líndal (1964b). This culminated a rapid growth during a 20 year period. By 1970 the greenhouse area was 110,000 m² as reported by Einarsson (1970). Thus, during the last decade the market for greenhouse products has apparently become saturated in Iceland and the greenhouse farmers have spent more time on developing better greenhouse farming methods than expanding their greenhouse area. About 70% of this area is used for vegetables, mostly tomatoes and cucumbers, giving an annual crop of about 1,000 tons, and about 30% of the area is used for flowers, such as roses, carnations, chrysanthemums and various potted plants. The total use of natural heat for this purpose in Iceland corresponds to some 20,000 tons of oil per year.

Greenhouses in Iceland are generally heated by means of radiator pipes which are mostly placed along the walls on the long side of the houses. A water temperature of 80° to 95 °C is common. Where it is hot enough, the water is also used to sterilize the soil after each season.

Dragone and Rumi (1970) describe the geothermal greenhouse developments in Japan. The most important products are horticultural. Flowers, such as lilies, chrysanthemums, carnations, orchids, and ornamental leaf plants are mainly cultivated. Potted plants are also produced. In some cases moderate amounts of melons, papayas, tomatoes, cucumbers and other fruits and vegetables are grown. Flowers, fruits and vegetables are cultivated mainly in winter, melons are cultivated rotationally during the season from spring to autumn. Papayas seem to be gathered at any time throughout the year. The bigger greenhouses in Japan are generally of the steel skeleton—glass type; smaller ones are often vinyl plastic houses.

Cooke (1970) describes a novel development in New Zealand, that of growing mushrooms by the use of natural heat. The soil is sterilized by the heat and the mushroom propagation houses are kept at the right temperature and humidity by using the geothermal fluid directly and without the use of heat exchangers. There is also a tree nursery, where seedlings are raised in greenhouses and shelters, which are used for thermal sterilization of soil.

Tikhonov and Dvorov (1970) and Sukharev et al. (1970) mention greenhouse developments in the U.S.S.R. Some are largely experimental, but many greenhouse centres produce important quantities of vegetables. Soil warming areas, where the soil may sometimes be covered with special materials, are frequently attached to the greenhouses. The greenhouses will operate all the year around, while soil warming allows a growing season from spring to autumn. Thus for instance in Makhach-Kala, thermal water has been heating greenhouses for several years. Two harvests of vegetables and flowers are obtained. A hotbed greenhouse centre of 50,000 m² is under construction combined with 50,000 m² of heated soil. At Grozni 100,000 m² of greenhouses will soon be added to the existing ones. In Kamchatka a 60,000 m² establishment has just been added to the older greenhouses. In the Caucasus several greenhouse establishments are mentioned.

These papers report that a single greenhouse generally covers 1,000 m² in the U.S.S.R. and that 15-20 kg of tomatoes and cucumbers may be produced per year from each m². It is mentioned that 40-50 kg of coal would otherwise be needed to grow 1 kg of tomatoes and cucumbers in greenhouses in all-year-round operation in the U.S.S.R.

In the U.S.A. Boersma (1970) reports that during 1969 field experiments were conducted near Corvallis, Oregon, designed to measure the effect of soil warming. The yield of corn silage increased by 45%, that of tomatoes by 50% (compared to usual out of doors growing in this area), soya bean silage by 66% and bus beans by 39%. He also reports a better quality of corn.

ANIMAL HUSBANDRY AND OTHER FARM USES

Boldizsar (1970) mentions a widespread use of geothermal water for animal husbandry in Hungary. It is used for heating cattle stalls, milking rooms, pigsties, chicken houses as well as for auxiliary rooms and the farm premises. Crop-drying at the farms is also reported.

Komagata et al. (1970) mention the use of geothermal water for animal husbandry in Japan, such as for hatching eggs and raising poultry, Sukharev et al. (1970) mention uses for dairy farming in the U.S.S.R.

In New Zealand geothermal energy is used to help biodegradation of the wastes from pigsties. The effectiveness of biodegradation helped by somewhat raised temperatures is discussed by Boersma (1970) from the standpoint of farming and the use which could be made of this product from big farms.

Washing and drying of wool with the aid of geothermal energy is practised.

In dairy farming there are many interesting opportunities for the application of natural heat other than those specifically mentioned above, such as the production of dried milk, casein and sucrose.

FISH HATCHING AND BREEDING

Matthiasson (1970) reports that several fish breeding stations in Iceland use geothermal water for tempering the water fed to the ponds. The largest is the Kollafjord

Experimental Fish Farm, where salmon are raised to the smolt stage, and either allowed to migrate to the sea from the farm or released in rivers in various parts of the country. About 300,000 smolts are produced per year. Water for the hatching cabinets is kept at 12°-14 °C by indirect heating by geothermal water. At least one breeding station for salmon smolts uses warm spring water directly.

Quoting Komagata et al. (1970), eels are bred in Hokkaido and Kagashima Prefectures in Japan, by utilizing heat from the hot springs. Eels bred in Ibusuki City of the latter Prefecture meet 50% of the total demand within the Prefecture. In Hokkaido Marine Hatching Centre the profitability of breeding eels has also been found to be promising. Geothermal water is mixed with river water and introduced to ponds where the temperature is kept at 23 °C. Seed eels are brought in from elsewhere, propagated and the eggs hatched. Finally, the eels are shipped as adult eels of 100 to 150 grams each after breeding for one to three years.

Alligators are also raised with the aid of geothermal water in Japan, at Atagawa Banana and Alligator Garden, Minami-Lzu-machi. Here, more than 20 kinds of alligators and crocodiles are bred in waste water at a temperature of 28°-32 °C. These are raised as a tourist attraction.

Recreational and health applications

Hot springs and warm mineral springs have been used for recreational and health purposes in many countries as long as historical records are available. Yet, a distinct change has occurred in recent years in that such places have become more accessible to many people, thanks to better communications and new hotels. There are records of very many geothermally heated swimming pools, mineral baths, mud baths, steam baths and specially organized recreational centres from several countries. But probably these applications of natural heat have been developed furthest in Japan, where hot springs have been used for this purpose from ancient times, and modern tourism has increased this impressively.

According to Komogata et al. (1970), a statistical survey made by the Welfare Ministry in Japan showed that in 1968 there were approximately 100 million visitors, which is equivalent to the total Japanese population. At the same time, statistics showed that about 20% of all sightseeing trips were for bathing, recreational and recuperative purposes. Including one-day visitors, not lodged, it is believed that about 150 million people visit hot springs annually in Japan.

Mashiko and Hirano (1970) explain that the main objects of this kind of hot spring utilization in Japan are for bathing, recreation and health in general; whereas medical treatment by doctors accounts for only a small proportion.

Geothermal energy in process heating

Perhaps the most striking characteristic of geothermal energy is its immense versatility. As a rule, its application in a major project will be governed by one or more specific objectives, but once it is introduced, certain other uses appear. Thus, the main objective may be some kind of evaporation, while the side uses may be drying, simple process heating, refrigeration, heating of industrial buildings or even the production of electrical power for the plant. Major industrial enterprises, and sometimes also agricultural ones, may become in effect combined schemes for the application of geothermal energy. Both of the independently developed major industrial schemes described above, that of the Tasman Company in New Zealand and that of Kísilidjan in Iceland, bear this out.

Another important consideration is the site selection for these enterprises. Unfortunately the price of geothermal energy increases sharply with the distance transmitted, so that the site should be chosen as close as possible to the source, but with due regard to all other considerations. In both the cases cited above, the plants are practically inside the geothermal areas. Good siting usually means that the most important factor in geothermal plant design, namely the low cost of the steam and water, has been established.

The cost of natural steam will of course vary from place to place, both because of geological conditions and local ones. Further, it will vary according to the length of the necessary collecting pipelines for a suitable plant site. But assuming reasonably favourable conditions, it is usually found that the cost of primary steam ranges from one-tenth to one-fifth of that of steam raised by fossil fuel. Further, flash steam, obtained from the water separated from the primary steam at higher pressures in the so-called wet geothermal fields, may sometimes be had at still lower cost, but of course this is low pressure steam.

The design of many appliances using steam is such that the investment is balanced against the cost of running the equipment. For example, in a multiple-effect evaporator the cost of steam will normally determine how many 'effects' are used. With high cost steam the 'effects' will be many and the investment relatively high in order to save steam; but with low cost steam the 'effects' will be few, and the investment almost proportionally lower. Design optimization of such a system, using geothermal steam, can frequently reduce the total production cost by more than the amount due to the difference in cost of geothermal and conventional steam alone.

The low cost of natural steam is in fact such a strong incentive that many new and more suitable processes may be established to take advantage of this, or older ones may be rearranged more or less completely. Yet, every such case has its own particular problems regarding raw materials, transportation, markets for the products and other circumstances, which must be thoroughly scrutinized before a recommendation can be made. A further consid-eration, dealt with briefly below, is the state of the source of energy as received.

THE STATE OF THE SOURCE OF ENERGY AS RECEIVED

With present exploitation techniques geothermal energy is received at the point of utilization as water at temperatures up to 100 °C or more, or as steam which may have satura-tion temperature as high as 200 °C.

At high reservoir temperatures there may be only steam from the boreholes or, more commonly, both steam and water. In the latter case, the reservoir temperature and the discharge pressure will determine the ratio of steam to water.

The water always contains dissolved solids, but their composition and concentration vary greatly. These solids may or may not interfere with the usefulness of the water as a source of energy, but always the first step in studying applications should be a chemical analysis, and an assess-ment of the chemical composition in respect to the service required.

For process heating, the scaling of heat transfer sur-faces and deposition of solids in the systems should be watched for. Occasionally, the water may also be unduly corrosive to ordinary structural materials.

The composition of these waters is very dependent on the reservoir conditions where they originate. Thus, the reservoir temperature will to a great extent determine their silica content and the soluble elements in the geological formation will determine many of the other characteristics.

Space does not permit a full discussion of these im-portant questions, but it may be said that where reservoir temperatures do not greatly exceed 100 °C, trouble with the deposition of siliceous materials is uncommon, while it is a major problem with water from high temperature reservoirs. Reference is made here to Hermannsson (1970) who studied scaling and corrosion in the Reykjavik heating system, where reservoir temperatures are around 90-130 °C. This system has been operating for about 30 years with very satisfactory results. The water is to a large extent used directly. Also, Yamagase et al. (1970), report on operations in Japan aimed at the reduction of scale formation from geothermal water originating in a high temperature reser-voir. The rate of scale formation is greatly reduced by pre-cipitation of silica in specially designed retaining channels.

The primary step to avoid corrosion is to exclude atmospheric oxygen from high temperature waters. High chloride contents will also enchance corrosion of carbon steels.

Natural steam may be received either dry saturated, superheated, or wet at the saturation temperature. Pressure may range from slightly above atmospheric to 15 to 20 ata.

To prevent scaling and the deposition of solids in process heating, it is usually advisable to wash dry steam with oxygen-free water or an alkaline solution. For one thing dry steam will frequently carry some solids which will deposit in the heating systems, and also some volatile materials may have to be washed out. Reference may be

145

made to Ozawa and Fujii (1970) who have discussed scale formation from superheated dry steam in Japan and to Allegrini and Benvenuti (1970) who have treated corrosion and scaling with dry steam in Italy.

When the steam is wet, the water usually must be separated anyway. This separation should be as complete as possible and the water will carry away the greater part of the solids and scale-forming elements.

Experience with process heating by wet steam has been reported in New Zealand and Iceland. In both cases scaling and corrosion problems are within acceptable industrial limits. In New Zealand this is mentioned by Smith (1970), who reports that in the Tasman Pulp and Paper Mill no trouble has been experienced with corrosion in the absence of oxygen, and that carbon steels have proved satisfactory; but such items as gland studs, valve seats and spindles are usually made of stainless steels. In this case the steam contains about $2\frac{1}{2}\%$ by weight of non-condensable gases (about 10 l/kg) of which 91% is CO_2 and the remainder mostly H_2S.

Experience with using natural steam in rotating steam tube dryers in the diatomite plant in Iceland is reported by Líndal (1970b). Here satisfactory results have been obtained by the use of mild carbon steel for the steam tubes. The same material is used for everything else in contact with the steam except parts where wear is possible, as in parts of the valves, steam traps, etc.: for these stainless steel is used.

Tolivia (1970) reports on corrosion tests which were made in a wet geothermal field in Mexico. He recommends that stainless steels with chrome content higher than 10% be used for those parts of equipment that need high resistance to corrosion; further, that it is advisable to use aluminium in equipment which needs corrosion resistance but does not justify the cost of stainless steel. It is advisable to prevent any mixture of air with the steam, since this greatly enhances the rate of corrosion; also to separate water as completely as possible from the steam, in order to minimize the presence of wetness and contaminants like chlorides (Tolivia, 1970).

Natural steam always contains some non-condensable gases. When this steam is used in process heating, these gases must be vented since their accumulation will decrease the overall heat transfer coefficient. It is therefore customary to release a few percent of the steam along with the gases in order to maintain satisfactory heat transfer conditions.

There are cases where the impurities of the geothermal fluids are objectionable. This is especially true for operations which need direct injection of steam or water with the processed materials or if the equipment is too sensitive to these impurities. In such cases the primary geothermal fluid may be used to produce a clean fluid by appropriate converters. This is occasionally done with both steam and water.

THE TEMPERATURE REQUIREMENTS OF DIFFERENT
APPLICATIONS

In process heating, practically the whole range of temperatures of geothermal fluids, both steam and water, may be utilised in one way or another; but specific applications often require definite geothermal feed temperatures, and either steam or water may be desirable.

In process design it is especially the minimum temperature requirements which often become a limiting factor. Too high a temperature can usually be adjusted without much effort.

In Figure 4 some applications have been arranged against a scale of temperature of geothermal fluids. This is intended as a rough illustration of the temperature requirements of different applications but it is emphasized that many of the items shown are practised with a range of feed temperatures rather than a single figure.

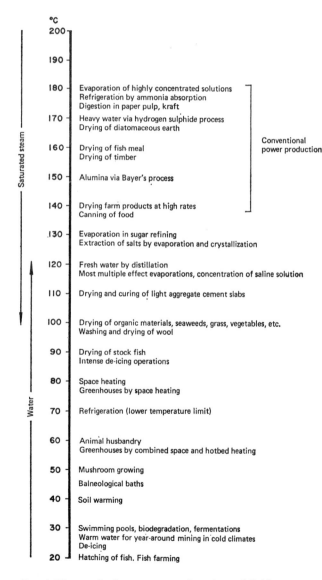

FIG. 4. The required temperature of geothermal fluids (approximate).

At temperatures above the boiling point of water, steam is often preferred while below this temperature water suits most applications. Apparently industrial applications mostly need the higher temperatures while space-heating and agriculture predominantly use low temperatures.

FUTURE DEVELOPMENTS

As with other things, the application of geothermal energy has both its problems and its rewards. Yet, experience shows that the problems have a way of disappearing as research and development advances in any specific application. But there are not many general solutions to the problems. It is much more likely that each new project will involve many new aspects more or less specific to itself. Hence, any widespread use of geothermal energy is very different on research and process development. This is especially true of industrial applications, because the differences lie not only in geothermal and local conditions, but also in the reevaluation or redesign of a great many industrial processes. However, the rewards should be such that there will be in the near future a flourishing activity.

Bibliography

Note. The reference to 'Pisa' signifies the United Nations Symposium on the Development and Utilization of Geothermal Resources, held in Pisa, 1970, the proceedings of which have been published by the Istituto Internazionale per le Ricerche Geotermiche, Lungarno Pacinotti 55, Pisa, Italy.

The reference to 'Rome' signifies the United Nations Conference on New Sources of Energy, held in Rome, 1961, the proceedings of which were published in 1964 by the United Nations, New York.

ALLEGRINI, G.; BENVENUTI, G. 1970. Corrosion Characteristics and Geothermal Power Production. Pisa IV/5.

ARMSTEAD, H. C. H. 1969. Geothermal Heat Costs. *Energy International, Brussels,* Febr. 1969, p. 28.

BODVARSSON, G. 1961. Utilization of Geothermal Energy for heating purposes and combined schemes involving power generation, heating and/or by-products. Rome GR/5 (G).

BOERSMA, L. 1970. Warm Water Utilization. *Conf. on Beneficial Uses of Thermal Discharges,* New York State Dept. of Conservation Environmental.

BOLDIZSAR, T. 1970. Geothermal Energy Production from Porous Sediments. Pisa I/3.

BURROWS, W. 1970. Geothermal Energy Resources for Heating and Associated Applications in Rotorua and Surrounding Areas. Pisa X/3.

CHILTON, C. 1953. *Process Requirements. Data and Methods of Cost Estimation.* Part. I. New York, McGraw Hill. 311 p.

COOKE, W. L. 1970. Some methods of Dealing with Low Enthalpy Water in the Rotorua Area of New Zealand. Pisa X/2.

DOWGIALLO, J. 1970. Occurrence and Utilization of Thermal Waters in Poland. Pisa III/3.

DRAGONE, G.; RUMI, O. 1970. Pilot Greenhouse for the Utilization of Low Temperature Waters. Pisa X/9.

EINARSSON, S. 1970. Utilization of Low Enthalpy Water for Space Heating, Industrial, Agricultural and Other Uses. Pisa X.

FACCA, G. 1970. General Report on the Status of World Geothermal Development. Pisa I.

FOSTER, P. K.; MARSHALL, T.; TOMBS, A. 1961. Corrosion Investigation in Hydrothermal Water at Wairakei, New Zealand. Rome G/47.

GARLADO, C. 1961. Problèmes techniques et économiques soulevés par la présence d'impuretés chimiques dans les fluides d'origine géothermique. Rome G/63.

HALLSSON, S. 1970. Drying Seaweeds and Grass by Geothermal Energy. *Tímarit Verkfraed. Islands,* no. 4.

HAYASHIDA, I.; EZIMA, Y. 1970. Development of Otake Geothermal Field. Pisa II/2.

HERMANNSSON, S. 1970. Corrosion of Metals and the Forming of Protective coating on the Inside of Pipes in the Thermal Water Used by the Reykjavik Municipal District Heating System. Pisa VIII/10.

ILUKOR, J. O. 1970. Geothermal Production of Electric Energy and Certain Minerals. Pisa II/7.

KOMOGATA, S.; IGA, H.; NAKAMURA, H.; MINOKARA, Y. 1970. The Status of Geothermal Utilization in Japan. Pisa II/3.

LENZI, D. 1961. Utilisation de l'énergie géothermique pour la production de l'acide borique et des sous-produits contenus dans les 'soffione'. Rome G/39.

LÍNDAL, B. 1961a. Geothermal Heating for Industrial Purposes in Iceland. Rome G/59.

——. 1961b. Greenhouses by Geothermal Heating in Iceland. Rome G/32.

——. 1970a. The Production of Chemicals from Brine and Seawater Using Geothermal Energy. Pisa V/i.

——. 1970b. The Use of Natural Steam in a Diatomite Plant. Pisa IX/11.

LUDVIKSSON, V. 1970. *Nýting jardhitans* (The Application of Natural Heat). Reykjavik, The National Research Council (Report 70-3). (In Icelandic.)

MAKARENKO, F. A.; MAVRITSKY, B. F.; LOKSHIN, B. A.; KONONOV, V. I. 1970. Geothermal Resources of the U.S.S.R. and Prospects for their Practical Use. Pisa II/6.

MASHIKO, Y.; HIRANO, Y. 1970. New Supply Systems of Thermal Springs to a Wide Area in Japan. Pisa X/6.

MATTHIASSON, M. 1970. Beneficial Uses of Heat in Iceland. *Conf. on Beneficial Uses of Thermal Discharges,* New York State Dept of Conservation Environmental.

MAZZONI, A. 1948. *The Steam Vents of Tuscany and the Larderello Plant,* p. 59-75. Bologna, Amonina Arts Grafiche. (In English.)

NAYMANOV, O. S. 1970. A Pilot Geothermoelectric Power Station in Pauzhetka, Kamchatka. Pisa II/14.

OZAWA, T.; FUJII, Y. 1970. A Phenomenon of Scaling in Production Wells the Geothermal Power Plant in the Matsukava Area. Pisa VIII/18.

PALMASON, G.; ZOËGA, J. 1970. Geothermal Energy Development in Iceland 1960-1969. Pisa II/20.

RAGNARS, K.; SAEMUNDSSON, K.; BENEDIKTSSON, S.; EINARSSON, S. 1970. Development of the Námafjall Area, Northern Iceland. Pisa IX/17.

SMITH, J. H. 1970. Geothermal Development in New Zealand. Pisa II/24.

SHCHERBAKOW, A. V.; DVOROV, V. I. 1970. Thermal Waters as a Source for Extraction of Chemicals. Pisa V/10.

SHREVE, R. N. 1956. *Chemical Process Industries* (2nd Ed.). New York, McGraw Hill.

SUKHAREV, G. M.; VLASOVA, S. P.; TARAMUKHA, Y. K. 1970. The Utilization of Thermal Waters of the Developed Oil Deposits of the Caucasus. Pisa I/26.

TIKHONOV, A. N.; DVOROV, I. M. 1970. Development of Research and Utilization of Geothermal Resources in the U.S.S.R. Pisa I/25.

TOLIVIA, M. E. 1970. Corrosion Measurements in a Geothermal Environment. Pisa VIII/12.

VALFELLS, A. 1970. Heavy Water Production with Geothermal Energy. Pisa IX/2.

WERNER, H. H. 1970. Contribution to the Mineral Extraction from Supersaturated Geothermal Brines, Salton Sea Area, California. Pisa.

WONG, C. M. 1970. Geothermal Energy and Desalination—Partners in Progress. Pisa.

YAMAGASE, T.; SUGINOHRA, Y.; YAMAGASE, K. 1970. The Properties of Scales and Methods to Prevent them. Pisa VIII/7.

V   Miscellaneous

Corrosion control in geothermal systems

T. Marshall and W. R. Braithwaite

Chemistry Division,
Department of Scientific and
Industrial Research (New Zealand)

Introduction

The natural geothermal steam and high-temperature water found in volcanic areas are characteristically contaminated with chemical impurities of underground origin, and during their utilization they may be further contaminated with impurities from the atmosphere. These impurities introduce corrosion problems which must be controlled in the design and operation of geothermal plants; furthermore the release of these impurities to the surface environment may introduce associated atmospheric and surface water corrosion problems.

The most common impurities encountered in geothermal fluids are:

Silica	Sodium
Chloride	Potassium
Fluoride	Lithium
Borate	Calcium
Sulphate	Magnesium
Carbonate	Ammonium

Hydrogen sulphide
Carbon dioxide
Hydrogen chloride

Of these the non-gaseous impurities, such as sodium chloride, are usually removed by separation (Marshall and Hugil, 1957) and/or scrubbing in the water phase before utilization of the steam phase in power generation (Haldane and Armstead, 1962) or heating systems (Sigurdsson, 1964). The gaseous impurities such as H_2S remain substantially in the steam phase (Marshall and Hugil, 1957), usually accompanied by residual traces of the non-gaseous impurities. After utilization and/or condensation of the steam the gaseous impurities become concentrated, contaminated with atmospheric oxygen, and may be released to the atmosphere.

Thus the non-gaseous impurities are usually of major significance in water-phase corrosion in geothermal systems, while the gaseous impurities are usually of major significance in steam-phase, condensate and atmospheric corrosion.

The many possible interactions between these chemical factors, physical factors such as temperature and stress, and the various materials of construction present the main problems of corrosion control in the design and operation of systems for the utilization of geothermal fluids.

Corrosion phenomena encountered in geothermal systems

The basic corrosion phenomena encountered in plant utilizing geothermal fluids may be discussed under various headings, as follows:

SURFACE CORROSION

Surface corrosion attack resulting in general surface wastage or pitting, usually of metal or concrete surfaces, has been investigated in New Zealand (Marshall and Hugil, 1957; Foster et al., 1964; Foster and Tombs, 1962), U.S.A. (Bruce and Albritton, 1959; Bruce, 1961) and other countries (Einarsson, 1964; Saporiti, 1964a), by direct engineering measurements, use of the A.S.T.M. coupon method (1946) and by corrosometer probe techniques (Dravnieks and Cataldi, 1954).

Surface corrosion may be extremely severe in geothermal fluids containing free hydrochloric, sulphuric or hydrofluoric acid (Tombs, 1960[1]) prohibiting the practical use of such fluids. Luckily most of the significant geothermal energy resources explored to date are not so highly contaminated with mineral acids; and, provided the fluids are not contaminated with atmospheric oxygen, surface corrosion rates of common structural materials are sufficiently low to permit their practical use.

Detailed surface corrosion rate data have been published by various investigators (Marshall and Hugil, 1957; Foster et al., 1964; Foster and Tombs, 1962; Bruce and

---

1. Unpublished investigations in New Zealand geothermal fields.

Albritton, 1959; Tombs, 1960) and are summarized in Table 1. The following general conclusions may be drawn:

(i) In air-free geothermal steam and high-temperature water, corrosion rates of the common engineering alloys are usually higher than those encountered in clean boiler-plant steam and water under similar temperature and pressure conditions;

(ii) Corrosion rates of most common engineering alloys in air-free geothermal fluids, with the possible exception of copper-base alloys, are low enough for their practical use in the construction of geothermal plants;

(iii) Aeration of geothermal media drastically accelerates the corrosion rate of most engineering alloys, with the notable and useful exceptions of austenitic stainless steel, titanium, and chromium (plating). The depolarizing action of oxygen introduced by aeration offers an obvious explanation of this acceleration;

(iv) Little published information is available on the corrosion performance of non-metallic materials in geothermal media, though more tests have been reported

(Sigurdsson, 1964; Bruce and Albritton, 1959; Bruce, 1964; Einarsson, 1964; Foster and Tombs, 1962). Concrete and grout are widely and satisfactorily used, and perform well except under conditions where atmospheric oxidation of H_2S can produce 'sulphate attack'.[1] Epoxy surface coatings on steel have also provided adequate protection against corrosion by aerated geothermal media (Einarsson, 1964);

(v) Surface corrosion of carbon steel and galvanized steel becomes severe in atmospheres contaminated by saline spray from geothermal bores, requiring the use of aluminium, stainless steel, or protective coating if the contamination cannot be avoided. Surface tarnishing of copper and silver is also very rapid in atmospheres contaminated by H_2S (Marshall, 1963) and may become of consequence in electrical, telephonic, and building rainwater equipment. Aluminium is not tarnished by H_2S.

1. Unpublished investigations in New Zealand geothermal fields.

TABLE 1. Surface corrosion rates of metals in geothermal media

Metal	Bore water[2] > 200 ºC	Water[3] ~ 125 ºC	Steam[4] 100-200 ºC	Aerated steam[5] ~ 100 ºC	Condensate[6] ~ 70 ºC	Condensate/ fresh water mixture[7] ~ 50 ºC	Highly acid thermal water[8]
Titanium	0	0	0	0	—	—	0
Chromium (plating on steel)	0		0	0	—	—	—
Aluminium	I	0.8-P	0-P-I	0-P	0.2	9	28
Zinc (coating on steel)	S[14]	1	0-I-P	S	—	S	—
Austenitic stainless steels[9]	0.1	0	0	0	0	0	22
Ferritic stainless steels[10]	0-0.1	0.1-P	0-0.3-P	1-P	0.1-P	0-0.5	—
Carbon and low alloy steels	0.3-0.4	0.3-0.5	0.3-6	20	3	30-170	1,000
Grey cast iron	1	0.4	1-3	10	—	90	—
High silicon cast iron	—	—	0.5	1	—	—	8
Brasses[11]	5	0.3	0.3-0.6	40	0.2	—	—
Bronze	20	—	2	9	—	—	—
Aluminium bronzes	10	—	2-3	10	1	—	—
Silicon bronze	—	—	3	20	—	—	—
Cupronickel	9	—	2	—	—	—	—
Beryllium copper	10	—	4	—	—	—	—
Copper	20	10	2	40	5	—	—
Nickel	6	—	1	8	2	—	—
Monel and K Monel	8-10	1	2-4	10	4	—	14
Nimonic 75	0.3	—	0	—	—	—	—
Inconel	1	0	0-0.3	80	—	—	20
Lead, antimonial lead	—	—	0.5	2.5-P	—	1	6

1. 1 mil = 0.001 inch. Data mainly from references 1, 4, 5 and 12.
2. Tests in water at bottom of a closed geothermal bore.
3. Water separated from wet geothermal steam at wellhead.
4. Steam separated from discharging geothermal bore.
5. Geothermal steam mixed with injected air.
6. Geothermal steam separated and condensed under pressure.
7. Geothermal steam condensed with freshwater to stimulate fluid in a jet condenser hot well.

8. Natural water in a volcanic crater (Tombs, 1960).
9. 18/8 CrNi, 18/8/3 CrNiMo, and 18/12/2 CrNiMo varieties.
10. 13 Cr, 17 Cr, 17/2 CrNi varieties.
11. 60/40 CuZn, arsenical 70/30 CuZn varieties.
I = internal attack with embrittlement.
P = pitting.
S = zinc coating stripped.

Surface corrosion information of this type (Table 1) can be usefully applied in selecting materials for specific items of equipment to utilize geothermal fluids, and in indicating operating precautions needed to control corrosion. For example, prevention of air leakage into low-pressure turbines, the use of protective coatings or resistant alloys (stainless steel, titanium) in jet condenser bodies and condenser gas extractors where aeration is unavoidable, and precautions against standby corrosion (Highley and Schnarrenberger, 1952) in shut-down plant, are usually necessary to minimize surface corrosion damage.

Mechanisms of surface corrosion of metals in geothermal media are discussed in more detail by Foster (1959).

EROSION-CORROSION

The conjoint action of erosion and corrosion (so-called erosion-corrosion) on metals is significant in some items of geothermal plant. Erosion-corrosion of turbine blades by wet steam at high velocity is particularly important in affecting the design and efficiency of steam turbines.

Empirical tests in geothermal steam (Marshall and Hugil, 1957) have shown that 13% Cr stainless steel blading alloys possess adequate erosion-corrosion resistance for geothermal steam service at 9% wetness and 900 fps. They have also shown that under these conditions erosion-corrosion resistance of metals is directly related to their static corrosion resistance in the same media, and not to hardness as is normally assumed. Thus improved resistance to erosion-corrosion would be expected from the more corrosion-resistant alloys such as austenitic stainless steels and titanium.

A more unusual variety of erosion-corrosion may be encountered in electrical commutators where tarnishing of the copper surfaces by atmospheric H_2S contamination from geothermal steam may accelerate 'wear' of the commutator surface at the zone of contact with the carbon brushes. The elimination of H_2S contamination reduces commutator wear to normal rates.

STRESS CORROSION AND SULPHIDE STRESS CRACKING

(a) *Stress corrosion of austenitic stainless steel*

Numerous investigators (Scheil, 1945; Copson, 1948) have reported stress corrosion of austenitic stainless steel in hot chloride solutions, usually concentrated solutions above 100 °C and under conditions where oxygen was not deliberately excluded. From laboratory tests and plant experience 5 ppm of chlorides and 50 °C have been suggested as minimum requirements for stress corrosion (Collins, 1955); there appears to be no limiting stress below which stress corrosion cracking will not occur.

Several investigators have reported this type of cracking in wet, chloride-contaminated steam (Edelnau and Snowden, 1957) and in geothermal steam and high-temperature water (Marshall, 1958; Neumann and Griess, 1963). Research in New Zealand (Marshall, 1958) has shown that, even under severe applied stresses, austenitic stainless steels are not susceptible to stress corrosion in air-free geothermal fluids, and that stress corrosion requires the presence of oxygen in addition to the other necessary factors, i.e. the presence of chloride solutions at high temperature, and tensile stress. Hoar and Hines (1956), Uhlig (Uhlig and Lincoln, 1958) Williams and Eckel (1956) have also shown that dissolved oxygen is essential for stress corrosion of austenitic stainless steels in chloride solutions.

This circumstance is of considerable practical importance in permitting the safe use of austenitic stainless steels in air-free geothermal fluids, where their high resistance to surface corrosion may be a great advantage, and in warning of the danger of their use in situations where aerated geothermal fluids are present at high temperature.

(b) *Stress corrosion of non-ferrous alloys*

Various investigators (Foster *et al.*, 1964; Foster and Tombs, 1962) have reported stress corrosion cracking of some non-ferrous alloys in geothermal fluids (Table 2). These alloys are of minor importance in geothermal plant and can be replaced by alternative resistant alloys.

(c) *Environmental stress cracking of non-metallic materials*

Embrittlement and cracking of plastic materials (usually known as environmental stress cracking (Howard, 1959), such as flexible PVC tubing, polyethylene, rubber, etc. has been experienced in New Zealand geothermal fields. The authors are aware of no systematic investigations of the stress cracking performance of plastic and non-metallic constructional materials in geothermal fluids.

(d) *Sulphide stress cracking*

Plant experience (Anon., 1952) and extensive laboratory research (Anon., 1952; Bastien, 1959a; Schuetz and Robertson, 1957) have shown that high-strength steels are susceptible to sulphide stress cracking, i.e. to spontaneous fracture when simultaneously stressed and exposed to aqueous solutions of H_2S. The literature contains abundant data on sulphide stress cracking in H_2S solutions below 100 °C, but much less information on the action of H_2S solutions at higher temperatures. Geothermal fluids frequently contain H_2S, hence it is not surprising that sulphide stress cracking in geothermal steam and high-temperature water has been reported by several investigators (Marshall and Hugil, 1957; Foster *et al.*, 1964; Foster and Tombs, 1962), mainly based on tests with constant-deformation stress-corrosion specimens (Champion, 1952). The interpretation of such tests is often made difficult by the occurrence of microscopic surface fissuring (Foster *et al.*, 1964) on specimens without complete macroscopic cracking. This fissuring is believed to represent a borderline condition of sulphide stress cracking or hydrogen-induced delayed fracture, probably related to the transient nature of hydrogen infusion as discussed below.

TABLE 2. Stress-corrosion behaviour of alloys in geothermal media

Alloy	Tensile strength, psi[1]	Stress corrosion behaviour in geothermal media[2]	
		Cracking[3]	Microfissuring[4]
Titanium	⩾ 93,000	No	No
Aluminium	16,000	No	No
Austenitic stainless steels	84-100,000	No	Aerated steam only
Ferritic[5] stainless steels	> 100,000	Yes	Yes
	< 100,000	No	No
Carbon and low alloy steels	⩾ 88,000	Yes	Yes
	< 88,000	No	Yes
Brass, 60/40	51-58,000	No	No
Brass, arsenical 70/30	54,000	Bore water only	No
Bronze	22,000	No	No
Aluminium bronzes	—	Yes	No
Silicon bronze	—	Yes	No
Cupronickel	90,000	No	No
Copper	36,000	No	Yes
Beryllium copper	Rockwell C38[6]	Yes	No
	Rockwell B47	No	Yes
Inconel	90,000	No	No
Nimonic 75	120,000	No	No
Monel	71-80,000	No	No
K Monel	Rockwell C25[6]	No	No

1. ⩾ indicates equal or greater than.
   < indicates less than.
   > indicates greater than.
2. Constant deformation test in media listed in Table 1. Data mainly from Marshall and Hugil (1957), Foster et al. (1964) and Foster and Tombs (1962).

3. Stress corrosion cracking in one or more of the above media.
4. Microscopic surface fissuring (Foster and Tombs, 1962), believed to be a borderline form of stress corrosion.
5. Martensitic in the hardened condition.
6. Hardened to spring temper.

Results of sulphide stress cracking investigations, mainly from tests in New Zealand geothermal media (Marshall and Hugil, 1957; Foster et al., 1964; Foster and Tombs, 1962) are summarized in Table 2. The following general conclusions may be drawn:

(i) Medium and high-strength carbon and alloy steels are susceptible to sulphide stress cracking under various environmental conditions in geothermal media up to at least 190 °C;

(ii) Low-strength carbon and alloy steels resist sulphide stress cracking in geothermal fluids. The strength level at which susceptibility to cracking occurs is not clearly defined and probably varies with type of steel, severity of stressing, and specific environmental conditions (see discussion below on delayed fracture). One investigation (Foster et al., 1964) reports that the threshold strength level for sulphide stress cracking appears to be about 88,000 psi for carbon and low-alloy steels,

and in the range 110,000-120,000 psi for high-chromium steels;

(iii) Sulphide stress cracking in geothermal fluids is influenced by physical condition of the specific environment, e.g. it was reported (Foster et al., 1964; Foster and Tombs, 1962) absent in geothermal water at 240 °C, even though it occurred in the same water cooled to 50 °C.

Sulphide stress cracking is a factor of major importance in the design and operation of equipment, particularly turbines, for the utilization of geothermal steam and water. The use of medium and high-strength steels is considered dangerous, on present knowledge, except in situations where the specific environment in known to be innocuous. The use of low-strength (less than 88,000 psi) steels is known to be safe for use in steam media under reasonable service stresses, both from test results and plant experience. These generalizations however must be treated with caution for condensate media where hydrogen-induced delayed fracture may occur as discussed below.

HYDROGEN INFUSION

Numerous investigators have shown that corrosion by aqueous solutions containing H_2S causes infusion of hygrogen into steels, as indicated qualitatively by hydrogen probe activity (Marsh, 1954). Hydrogen infusion is known, in favourable circumstances, to cause blistering and embrittlement of steels, and to be associated with sulphide stress cracking and delayed fracture of stressed steels.

In geothermal fluids blistering of steel appears to be observed only rarely (Foster et al., 1964), in grossly laminated steel. Detectible embrittlement of steel as measured by loss of elongation in conventional tensile tests has been reported to be measurable only after exposure of steel to geothermal steam condensate saturated with H_2S (Foster et al., 1964).

Hydrogen probe tests (Foster, 1962) have provided more positive evidence of hydrogen infusion. Results of a large number of carbon steel probe tests in geothermal fluids reported by Foster (1962) are summarized in Figure 1. The positions of the curves are not highly reproducible for any particular medium, but the curve shapes are characteristic for each medium and fall into the general patterns indicated. Several significant deductions can be drawn from these curves:

(i) In geothermal media above 100 °C the hydrogen permeation rates fall off with time to very low values after 2 to 3 weeks, probably due to the development of external corrosion products on the probe bodies and reduction of external corrosion rates. This decrease in permeation rate explains, at least qualitatively, why blistering and embrittlement do not occur readily in geothermal media above 100 °C;

(ii) The total quantity of hydrogen permeated through steel in a given time can be greatly reduced by surface coatings such as 'Apexior' paint, oxidation, and chemically formed magnetite coatings;

(iii) Aeration of geothermal media either eliminates or drastically reduces hydrogen permeation. The cathodic depolarizing action of dissolved oxygen offers an obvious explanation of this effect;

(iv) A marked inverse relation exists between temperature and hydrogen probe activity in geothermal media. Foster (1959) has shown that the corrosion product on carbon and low alloy steels consists of iron sulphide alone in cold condensate, of magnetite beneath iron sulphide in water/steam media from 100 to 180 psig, and of magnetite alone in bore water at 240 °C. This suggests strongly that the change in hydrogen permeation with temperature is related to changes in corrosion mechanism;

(v) The typical shapes of carbon steel probe activity curves in Figure 1 may be interpreted qualitatively as follows. If, as a first approximation, hydrogen permeation is assumed to follow Fick's law (which is not strictly

true — McNabb and Foster, 1963; Foster et al., 1965), the permeation curves imply (Foster, 1962) that steel immersed in cold condensate and low-pressure steam will gradually saturate with hydrogen to finite concentration levels, whereas steels immersed in high-pressure steam and high-temperature water will experience transient injection of hydrogen, especially in zones close to the steel surface, which will later diffuse out leaving very low hydrogen concentrations under equilibrium conditions. The measured equilibrium concentrations of hydrogen in steel exposed to these media (Marshall and Tombs, 1969) conform to this suggested mechanism.

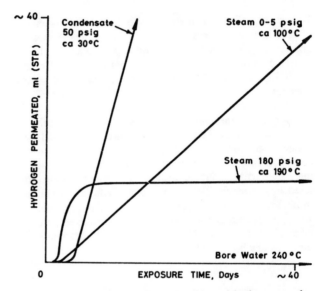

FIG. 1. Generalized behaviour of mild steel hydrogen probes exposed to geothermal media (Exposed surface area ~ 60 cm²).

The transient nature of hydrogen infusion offers one possible explanation of the formation of non-propagating microscopic surface fissuring observed in stress corrosion tests.

HYDROGEN-INDUCED DELAYED FRACTURE

Troiano (1960), Schuetz and Robertson (1957) and Bastien (1959) have shown that hydrogen infusion can produce delayed fracture in high-strength steels subjected to tensile stress. The hydrogen probe tests discussed above have also shown that hydrogen infusion results from exposure of steel to geothermal fluids. Hence delayed fracture of tensile-stressed high-strength steel would be expected during exposure to geothermal fluids. In fact the researches of many investigators suggest that sulphide stress cracking, delayed fracture, and hydrogen infusion are inter-related phenomena (Vollmer, 1958).

Marshall and Tombs (1969) have provided quantitative data on delayed fracture of bore casing steels exposed to geothermal fluids in New Zealand for long periods, under closely-controlled tensile loading. This work showed that no detectable delayed fracture occurred in geothermal steam media, and that both high-strength and low-strength casing steels exhibited delayed fracture when exposed to cold geothermal steam condensate. Of the casing steels tested, the French steel APS 10M4 (Cauchois *et al.*, 1957) exhibited maximum resistance to delayed fracture.

The results of these tests, which were of necessity a measure of stress corrosion as well as delayed fracture under quite precisely controlled stress distributions, suggest that it may be safe to use steels of higher strength than 88,000 psi in geothermal steam media (but not in condensate or water).

CORROSION-FATIGUE

Gilbert (1956) has drawn attention to the deleterious effects of simultaneous corrosion on the fatigue life of metals. Similar deleterious effects on fatigue properties would be expected from corrosion by geothermal media, and could be of significance in the design and operation of steam turbines.

No quantitative corrosion-fatigue data for metals exposed to geothermal media are known to the authors, but some relevant data in the literature (Westcott, 1938; Westcott and Bowers, 1933) indicate that unusually severe corrosion fatigue is encountered in salt solutions containing dissolved H_2S, similar to fluids which may be encountered in geothermal fields. Experience with fatigue failure of blades in turbines operated on New Zealand geothermal steam suggests that corrosion fatigue is operative, but can be controlled by careful turbine design. Quantitative investigations are being undertaken.

Corrosion control in design and operation of geothermal systems

Corrosion problems encountered in specific items of plant using geothermal fluids in various countries, and methods used for their control, may be discussed under the following broad sectional headings.

BORES AND WELLHEAD EQUIPMENT

Conventional low-strength casing steels, such as API grades H40 and J55, are commonly used with satisfactory performance in geothermal bores (Fooks, 1964). Corrosion is usually insignificant due to absence of oxygen in geothermal bore fluids; however the presence of aerated thermal groundwater close to the surface can produce significant external corrosion at the surface end of casing strings. This is usually controlled by the use of multiple casings at the surface with careful grouting of the annuli. Accidental deflection of the high velocity steam/water/detritus stream flowing during initial discharge of bores has caused dramatic erosion-corrosion damage to casings and valves

(Smith, 1964), but erosion-corrosion usually presents no unusual problems in casings during normal operation.

The severe thermal stresses resulting from intermittent flowing of geothermal bores can produce casing fractures and joint failures (Smith, 1964). It is not known whether sulphide stress cracking and delayed fracture are involved in these failures, due to the virtual impossibility of recovering broken strings undamaged for examination. Research on delayed fracture (Marshall and Tombs, 1969) discussed above suggests, however, that this failure of low strength steel casing strings probably occurs by mechanical action, whereas medium and high-strength strings would be more likely to fail by delayed fracture mechanisms. Hence N80 and higher strength casing steels are considered, on present evidence, to be susceptible to delayed failure, with the possible exception of the French steel APS 10M4 specially designed to resist sulphide cracking and hydrogen-induced delayed fracture. The danger of casing fractures from mechanical, stress corrosion and delayed fracture causes, can be minimised if not eliminated by keeping bores discharging or hot after initial 'blowing in'.

Wellhead equipment is usually constructed of low-strength carbon steels, with stainless steels for valve trim, which perform satisfactorily. Copper-base alloys should be avoided because of their low corrosion resistance, though brass often shows sufficient resistance for use in minor equipment details. Maintenance problems may arise where air and leaking or discharging fluids mix; for example, valve packing leakage can cause spindle corrosion and seizure, hence attention to tight packing is required. Atmospheric pollution with discharged steam or water can cause severe corrosion of cold unprotected steel, pressure gauges, control equipment, etc., requiring protection of such equipment by painting, cladding, etc. or avoidance of the pollution by remote discharge of waste fluids down-wind of equipment. Any stressed stainless steel equipment must be designed to avoid contamination by hot aerated steam or spray to prevent chloride stress cracking. Discharge of high-velocity wet steam, e.g. in silencers (Dench, 1964) can produce rapid erosion of deflector plates. This is minimized by attention to streamline flow in design, or by use of non-metallic materials.

STEAM/WATER PIPELINES

Cross-country pipelines for transmission of geothermal steam, steam-water mixtures, and water (Haldane and Armstead, 1962; Sigurdsson, 1964) from wellheads to power stations, constructed in low-strength carbon steels, have provided satisfactory corrosion-free service. The main corrosion problem arises from design methods used to cope with thermal expansion. Expansion loops introduce no corrosion control demands, but the use of bellows-type expansion compensators requires care in design and operation. Austenitic stainless steels are sufficiently resistant to surface corrosion for such service; however, due to the high operating temperature and high operational stresses, chloride stress cracking can occur very rapidly if external surfaces of the bellows are exposed to chloride-contaminated steam, water or spray. Steam leakage into the bellows must be strictly avoided. With suitable design and operational precautions austenitic stainless steel compensators have given excellent service in New Zealand geothermal installations (Marshall, 1963).[1]

The main operating precaution in such pipelines is the avoidance of standby corrosion (Marshall, 1963; Highley and Schnarrenberger, 1952). If pipelines are shut down without such precautions, the ingress of air during cooling, with consequent oxidation of residual H_2S, produces highly acid condensate which can cause serious pitting of stainless and carbon steel equipment. The simplest standby procedure is usually to keep such pipelines full of steam to avoid air ingress during standby periods. Design of pipeline systems must essentially include design for control of standby corrosion.

TURBINES

When heat exchangers (constructed of low-strength carbon steel) are used to generate clean steam for turbine operation (Hansen, 1964; Zancani, 1964) no abnormal corrosion factors are involved in turbine design and operation. When direct geothermal steam feed to turbines is used, as in Italy (Zancani, 1964; Saporiti, 1964b; Villa, 1964), U.S.A. (Bruce and Albritton, 1959; Bruce, 1964), New Zealand (Haldane and Armstead, 1962), precautions are required to cope with the corrosion phenomena discussed above.

The high speeds and stress levels of turbine rotary components, and the catastrophic nature of rotor failures (Thum, 1956-1961) lead to strict design precautions to avoid rotor and blade failure. From the corrosion viewpoint these usually include the use of only low-strength carbon and alloy steels to avoid the danger of sulphide stress cracking, and sealing of joints in low-pressure turbine casings as thoroughly as possible to minimise air leakage into the casing with consequent acceleration of internal corrosion. With these precautions, conventional turbine constructural materials (cast steel casings, carbon steel rotors, 13% Cr steel blades, shrouds and lacings heat-treated to the low-strength condition) have given excellent service in Italy (Saporiti, 1964b), U.S.A. (Bruce and Albritton, 1959; Bruce, 1964; Hansen, 1964), and New Zealand (Haldane and Armstead, 1962). In the latter installation mechanical attachment of lacings and shrouds was used instead of brazing, to avoid the danger of local hardening of 13% Cr steel during brazing operations. Shroud rivets have been annealed carefully to minimise residual stresses. Maximum blade tip velocities and steam wetness were also limited to 900 fps and about 9% respectively to avoid excessive blade erosion without recourse to erosion shields brazed into the blades.

These limitations to the use of low-strength steels and relatively low blade-tip velocities have significant effects on geothermal steam turbine design. Within these limitations the use of 1,500 rpm rotor speed has permitted the generation of 30 MW in single-exhaust geothermal steam turbines in New Zealand (Haldane and Armstead, 1962). From available published information (Saporiti, 1964b; Kearton, 1951), similar principles and limitations appear to be involved in Italian and U.S.A. geothermal steam turbine installations.

Conventional turbines of the largest sizes (with higher power outputs per exhaust) may require medium strength steels to cope with operating stresses in the final stages, together with erosion shields brazed, or in some cases welded, on to blades in the final stages to prevent erosion at high blade-tip velocities. It is not yet known from operating experience whether these principles can be applied in geothermal plants to improve economics by the use of machines with a larger power output per exhaust. The present scientific evidence on delayed fracture and stress corrosion cracking of medium and high-strength steels in geothermal steams (see above) is rather conflicting. The delayed fracture investigations discussed above suggest that some medium- and high-strength steels are not susceptible to delayed fracture and stress corrosion when exposed to steam, as distinct from condensate. However, the effects of wet steam at high velocity have not been assessed, and would be expected to enhance hydrogen infusion and the danger of delayed fracture. Also the effects of corrosion-fatigue on such materials are not quantitatively known. Further scientific investigation of these factors, plus materials testing in operating turbines, will be needed to determine whether higher strength steels and erosion shields can be used safely in the construction of higher powered geothermal turbine units. There is of course no inherent objection to the use of multiple exhausts for high-output turbo-generating sets in geothermal power plants.

Fatigue or corrosion-fatigue has been encountered in the blades of some geothermal steam turbines in New Zealand. Although mechanical factors were involved in these failures (Haldane and Armstead, 1962), their persistence suggested that blading steels are more susceptible to corrosion-fatigue in geothermal steam than in clean, boiler-plant steam. Quantitative corrosion-fatigue data for geothermal steam is notably lacking.

In geothermal turbines, labyrinth glands are items susceptible to corrosion, since at these points leaking steam meeting atmospheric oxygen can produce a highly corrosive environment. Design precautions include the use of corrosion-resistant gland materials (soft austenitic stainless steel, chrome iron), and pressure-feeding of the glands with clean steam from heat exchangers to prevent any egress of impure geothermal steam.

The main corrosion precautions in turbine operation are the normal attention to preventing standby corrosion (Haldane and Armstead, 1962) and care in joint sealing to prevent excessive air leakage into LP[2] sets operating below atmospheric pressure.

1. Unpublished investigations in New Zealand geothermal fields.
2. Low pressure.

CONDENSER SYSTEMS

In the condenser systems of geothermal power plants the mixing of geothermal steam or condensate with atmospheric oxygen is inevitable. With surface condensers air invariably leaks into the LP system; with jet condensers oxygen enters dissolved in the cooling water. Hence corrosion conditions attain maximum severity in the condenser system, and especially in the gas extraction system where wet H_2S, carbon dioxide and oxygen mixtures can react to produce highly acid corrosive media (Zancani, 1964; Saporiti, 1964*b*).

Corrosion control measures include the use of mild steel with epoxy and other proprietary surface protective coatings (Haldane and Armstead, 1962; Bruce, 1964) lead-coated steel (Zancani, 1964), 13% Cr steels where corrosion conditions are not a maximum severity, aluminium (Bruce, 1964), austenitic stainless steels (Bruce and Albritton, 1959; Bruce, 1964; Zancani, 1964), P.V.C.—polyester/fibreglass and other plastic materials (Bruce, 1964) and wood (Bruce, 1964) for highly corrosive areas where temperatures and pressures are relatively low, pyrex glass tubing, 'Niresist' iron for pumps, and concrete for water discharge ductings (Bruce, 1964). Austenitic stainless steel resists all but the most corrosive conditions (hot acid condensate at pH less than 4 causes rapid attack), but must be used with caution in equipment such as steam jet ejectors where slight super-heating may occur and cause rapid chloride stress cracking. This can usually be prevented by deliberate de-superheating.

Plastic materials and glass are usually limited to low-temperature, low-pressure parts of the condenser systems such as condensate drains and gas exhaust ducts; concrete water ducts may require surface protection (e.g. with coal tar coatings — Bruce and Albritton, 1959) to prevent 'sulphate attack' in vapour spaces above water level.[1]

Centrifugal compressors for gas extraction from condensers operate with a highly corrosive mixture of wet gases and air. A novel method of controlling corrosion in the compressor system is to limit the amount of interstage cooling so that the gases remain relatively hot, dry and therefore non-corrosive during their passage through the compressor system.[1]

COOLING TOWERS

Cooling towers serving jet condensers in geothermal power plants meet abnormal conditions, mainly caused by dissolution of H_2S in the circulating water and its subsequent oxidation in the cooling tower to sulphur, sulphuric acid, and other sulphur compounds. The circulating water may become 'milky' due to build-up of fine suspended sulphur. Corrosiveness of the circulating water appears to depend largely on the amount of ammonia (or ratio of ammonia to H_2S) in the steam, since ammonia also dissolves in the circulating water and tends to neutralize sulphuric acid formed from oxidation of the H_2S. Thus in the Geysers plant with $NH_3 : H_2S$ ratio about 0.3 : 1 the pH of the circulating water is about 7 (Bruce and Albritton, 1959); at Larde-rello with $NH_3 : H_2S$ ratio about 0.2 : 1 the pH of circulating water is also about 7;[2] while unpublished tests at Wairakei with $NH_3 : H_2S$ ratio less than 0.1 : 1 have shown that the circulating water can become highly acid with pH 3.0.

Corrosion control measures in cooling tower design include the use of wood, concrete, stainless steel, aluminium, asbestos board, plastics, and non-metallic protective coatings as corrosion resistant materials (Bruce and Albritton, 1959; Bruce, 1964).

AUXILIARY EQUIPMENT

The use of plastic piping buried in thermal ground for water reticulation and for the transport of geothermal fluids should be approached with caution since such materials can suffer embrittlement or environmental stress cracking in these media,[1] and they are often permeable to H_2S, resulting for example in contamination of the water conveyed in them. Selection of materials for underground piping may be very difficult since thermal ground is often highly corrosive to metals and concrete. At present the most practical system appears to be the use of metal or asbestos-cement piping protected with carefully applied non-metallic coatings.

The use of spring materials in control and measuring equipment exposed to geothermal steam (such as bourdon tubes in pressure gauges, recorders, etc.) often presents problems due to stress corrosion cracking of steel springs, corrosion and stress cracking of copper-based spring alloys such as beryllium copper. Preventive measures include the use of stainless steel and K monel spring materials, or the use of isolating fluids where practicable in pressure gauging, recording systems etc.

Atmospheric contamination with hydrogen sulphide, frequently prevalent in geothermal fields, can cause irritating and insidious corrosion problems associated with tarnishing of copper and silver alloys, and blackening of lead-pigmented paints. These effects are often unsightly rather than serious, e.g. blackening of lead paints, tarnishing of copper tubing, instrument cases, overhead power cables, rain-water ducting etc. In some cases however they can have serious effects, as when tarnishing of silver contacts makes telephone communications or electrical recording/controlling equipment inoperative, and when tarnishing of copper produces severe erosion-corrosion of copper switch contacts. Preventive measures include limitation of H_2S contamination (e.g. gas discharge through remote or high vents); air conditioning to remove H_2S (this method is usually expensive and insufficiently effective to protect silver contacts from tarnishing); preventive maintenance; and selection of alternative resistant materials. For example the use of aluminium for overhead conductors, electrical equipment and building sheathing, instead of copper, eliminates tarnishing; titanium oxide pigmented paints

---

1. Unpublished investigations in New Zealand geothermal fields.
2. Unpublished information from the Larderello geothermal field.

resist blackening; chromium plated instrument casings and components of telephones and gauges resist tarnishing; platinum gold, gold plating, and other noble contact alloys in preference to silver prevent tarnishing and defective operation of electrical instrument contacts. The multitude of electrical contacts in a modern power station make this aspect particularly important. Practice in New Zealand[1] involves the replacement of silver contacts in critical equipment with platinum contacts, and continuous preventive

maintenance by cleaning and lubrication in the case of non-critical equipment.

This often-ignored aspect of atmospheric contamination and tarnishing of copper and silver alloys is, in practice, one of the most troublesome geothermal corrosion problems which warrants full consideration in the plant-design stage.

1. Unpublished investigations in New Zealand geothermal fields.

Bibliography

ANON. 1952. Symposium on sylphide stress corrosion. *Corrosion*, 8, (10).

American Society for Testing Materials. 1958. *A.S.T.M. Standards 1958, Part 3. Recommended Practice for Conducting Plant Corrosion Tests*, p. 257-265. Philadelphia, American Society for Testing Materials (A.S.T.M. Designation A 224-46).

BASTIEN, P. G. 1959a. *Physical metallurgy of stress corrosion fracture*. New York, Interscience, 311 p.

——. 1959b. The phenomena of cracking and fracture of steel in the presence of hydrogen. *Physical metallurgy of stress corrosion fracture*, New York, Interscience, 311 p.

BRUCE, A. W. 1961. Experience of generating geothermal power at the Geysers power plant, Sonoma County, California, U.S.A. *U.N. Conf. on New Sources of Energy, Rome, 1961.* (E/Conf. 35/G/8) New York, United Nations.

——; ALBRITTON, B. C. 1959. Power from geothermal steam at the Geysers power plant. *Proc. Amer. Soc. civ. Engrs*, 2287 (J. Power Div.).

CAUCHOIS, L.; DIDIER, J.; HERZOG, E. 1957. A special N80 steel tubing developed in France to resist sulphide stress corrosion in sour gas wells. *Corrosion*, 13, 263t-269t.

CHAMPION, F. A. 1952. *Corrosion testing procedures*. London, Chapman & Hall.

COLLINS, J. A. 1955. Effect of design, fabrication, and installation on the performance of stainless steel equipment. *Corrosion*, 11, 11t-18t.

COPSON, H. R. 1948. Stress corrosion. In: H. H. Uhlig (ed.) *Corrosion Handbook*, p. 569-578. New York, Wiley.

DENCH, N. D. 1961. Silencers for geothermal bore discharge. *U.N. Conf. on New Sources of Energy, Rome, 1961.* (E/Conf. 35/G/18) New York, United Nations.

DRAVNIEKS, A.; CATALDI, H. A. 1954. Industrial applications of a method for measuring small amounts of corrosion without removal of corrosion products. *Corrosion*, 10, (7), 224-230.

EDELEANU, C.; SNOWDEN, P. P. 1957. Stress corrosion of austenitic stainless steels in steam & hot water systems. *J. Iron St. Inst. 186*, 406.

EINARSSON, S. S. 1961. Proposed 15 MW geothermal power station at Hveragerdi, Iceland. *U.N. Conf. on New Sources of Energy, Rome, 1961.* (E/Conf. 35/G/9.) New York, United Nations.

FOOKS, A. C. L. 1961. The development of casings for geothermal boreholes at Wairakei, New Zealand. *U.N. Conf. on New Sources of Energy, Rome, 1961.* (E/Conf. 35/G/16.) New York, United Nations.

FOSTER, P. K. 1959. The thermodynamic stability of iron and its compounds in hydrothermal media. *N.Z. J. Sci, 2*, (3), 422-430.

——. 1962. Some aspects of hydrogen infusion into steels exposed by hydrothermal media. *Aust. Corros. Engng, 6*, (10), 306.

——; MARSHALL, T.; TOMBS, A. 1961. Corrosion investigations in hydrothermal media at Wairakei, New Zealand. *U.N. Conf. on New Sources of Energy, Rome, 1961.* (E/Conf. 35/G/47.) New York, United Nations.

——; MCNABB, A.; PAYNE, C. M. 1965. On the rate of loss of hydrogen from cylinders of iron and steel. *Trans. Met. Soc. A.I.M.E., 233*, 1022-1031.

——; TOMBS, A. 1962. Corrosion by hydrothermal fluids. *N.Z. J. Sci. 5* (1), 28-42.

GILBERT, P. T. 1956. Corrosion fatigue. *Metallurg. Rev., 1*, 379.

HALDANE, T. G. N.; ARMSTEAD, H. C. H. 1962. The geothermal power development at Wairakei, New Zealand. *Proc. Instn. mech. Engrs., Lond. 176* (23) 603-648.

HANSEN, A. 1961. Thermal cycles for geothermal sites and turbine installation at the Geysers power plant California. *U.N. Conf. on New Sources of Energy, Rome, 1961.* (E/Conf. 35/G/41.) New York, United Nations.

HIGHLEY, L.; SCHNARRENBERGER, W. R. 1952. Prevention of standby corrosion in power plants. *Corrosion*, 8, 171.

HOAR, T. P.; HINES, J. G. 1956. The stress corrosion cracking of austenitic stainless steels. *J. Iron St. Inst. 182*, (2), 124; *184*, (2), 166.

HOWARD, J. B. 1959. A review of stress-cracking in polyethylene. *S.P.E.J., 15*, (5), 397-412.

KEARTON, W. J. 1951. *Steam turbine theory and practice* (6th Ed.), London, Pitman, 495 p.

MARSH, G. A. 1954. Some notes on hydrogen blistering. *Corrosion*, 10, 101.

MARSHALL, T. 1958. Stress corrosion of austenitic stainless steel in geothermal steam. *Corrosion, 14*, 159t-162t.

——. 1963. Corrosion control in geothermal power production. *Aust. Corros. Engng., 7*, 9-11.

——; HUGIL, A. J. 1957. Corrosion by low pressure geothermal steam. *Corrosion, 13*, 329t.

——; TOMBS, A. 1969. Delayed fracture of geothermal bore casing steels. *Aust. Corros. Engng. 13*, (9), 2-8.

MCNABB, A.; FOSTER, P. K. 1963. A new analysis of the diffusion of hydrogen in iron and ferritic steels. *Trans. Met. Soc. A.I.M.E., 227*, 618-626.

NEUMANN, P. D.; GRIESS, J. C. 1963. Stress corrosion cracking of type 347 stainless steel and other alloys in high temperature water. *Corrosion, 19*, 345t-333t.

SAPORITI, A. 1961a. Progress realized in installations with endogenous steam turbine-generator units without condenser. *U.N. Conf. on New Sources of Energy, Rome, 1961.* (E/Conf. 35/G/64.) New York, United Nations.

——. 1961b. Progress realized in installations with endogenous steam condensing turbine-generator units. *U.N. Conf. on New Sources of Energy, Rome, 1961.* (E/Conf. 35/G/60.) New York, United Nations.

SCHEIL, M. A. 1945. Some Observations of Stress-Corrosion Cracking in the Austenitic Stainless Alloys. *Symposium on Stress Corrosion Cracking of Metals, 1944,* p. 395-410. Published jointly by the American Society for Testing Materials, Philadelphia and the American Institute of Mining and Metallurgy Engineers, New York.

SCHUETZ, A. E.; ROBERTSON, W. D. 1957. Hydrogen absorption, embrittlement and fracture of steel. *Corrosion, 13,* 437t.

SIGURDSSON, H. 1961. Reykjavik Municipal District Heating Service and its experience in utilizing geothermal energy for domestic heating. *U.N. Conf. on New Sources of Energy, Rome, 1961.* (E/Conf. 35/G/45.) New York, United Nations.

SMITH, J. 1961. Casing failures in geothermal bores at Wairakei. *U.N. Conf. on New Sources of Energy, Rome, 1961.* (E/Conf. 35/G/44.) New York, United Nations.

THUM, E. E. 1956-1961. Recent accidents with large forgings. *Metal Progr. 69,* (2), 49-57, Feb. 1956. Further discussions, Ibid. *69,* (2), 63-67, April 1956; *76,* 72-75, July 1959; *79,* 89-91, August 1961.

TOMBS, A. 1960. Corrosion of metals in Whangaehu River and Ruapehu Crater waters. *N.Z. J. Sci., 3,* (1), 93-99.

TROIANO, A. R. 1960. The rôle of hydrogen and other interstitials in the mechanical behaviour of metals, *Trans. Amer. Soc. Metals, 52,* 54.

UHLIG, H. H.; LINCOLN, J. 1958. Chemical factors affecting stress corrosion cracking of 18-8 stainless steels. *Trans. electrochem. Soc. 105,* 325.

VILLA, F. 1961. Latest trends in the design of geothermal plants. *U.N. Conf. on New Sources of Energy, Rome, 1961.* (E/Conf. 35/G/72.) New York, United Nations.

VOLLMER, L. W. 1958. The behaviour of steels in hydrogen sulphide environments. *Corrosion, 14,* 324t-328t.

WESTCOTT, B. B. 1938. Fatigue and Corrosion-Fatigue of Steels. *Mech. Engng., N.Y.,* vol. 60, p. 813-822.

——; BOWERS, C. N. 1933. Explanation of the Mechanism of Corrosion Fatigue and its Application to Sucker Rod Failures. *Oil Gas J.,* vol. 32, no. 23, p. 65-69.

WILLIAMS, W.; ECKEL, J. F. 1956. Stress corrosion of austenitic stainless steels in high temperature water. *J. Amer. Soc. nav. Engrs., 68,* 93.

ZANCANI, C. F. A. 1961. Comparison between surface and jet condensers in the energetic and chemical utilization of Larderello's boraciferous steam jets. *U.N. Conf. on New Sources of Energy, Rome, 1961.* (E/Conf. 35/G/50.) New York, United Nations.

Geothermal economics

H. Christopher H. Armstead
Consulting Engineer
Rock House, Ridge Hill,
Dartmouth, South Devon
(United Kingdom)

1. General

For any 'heat-intensive' process such as space heating, power production, distillation and certain manufactures it is of the utmost importance that cheap heat should be available. Geothermal energy can provide one of the cheapest sources of heat in the world, without prejudice to the other cost components which together account for the cost of the end product; i.e. capital cost of the necessary plant and equipment, the cost of raw materials fed into the process, attendance and maintenance costs, etc.

It is of course well known that the cost of anything can usually be expressed in two terms—a *fixed* component which is incurred simply by establishing the production process and which is independent of the quantity of production, and a *variable* component which is directly proportional to the quantity of production. The cost of geothermal heat is incurred entirely in the form of fixed costs without any variable component whatsoever. Hence, geothermal heat is eminently suited to any process requiring a continuous, rather than an intermittent, supply of heat, so that the fixed costs may be spread over the greatest possible quantity of heat and the cost per calorie may be low.

The development of geothermal energy usually requires fairly heavy initial expenditure on exploration before the geothermal fluids start to flow in worthwhile quantities. If geothermal energy is to compete with fuel as a source of heat it must be clearly demonstrated to be cheaper than the alternative after bearing its full share of these exploration costs. Even where the costs of exploration have been wholly or partly borne by some international organisation, these costs are just as real and should be taken into account when assessing the cost of the heat won.

An attempt will now be made to assess representative costs applicable to a 'typical' developed field. Such an attempt can only establish the order of magnitude of geothermal heat costs, but the exercise is worthwhile in that it enables a reasonable standard to be set up, against which the actual costs in certain fields may be compared.

2. Cost components

The cost components that enter the reckoning of geothermal heat costs may be listed as follows:

2.1 CAPITAL COSTS

Exploration costs:

> Topographical and geological surveys, geophysical and geochemical investigations, exploratory drilling and field investigations.

Drilling costs:

> The sinking of production bores (including the costs of a certain proportion of bores which may prove to be failures) at locations which appear to have the greatest probability of achieving high production, as determined by the results of exploration.

Wellhead equipment costs:

> The supply and installation of suitable valving, separators, silencers and instrumentation, with incidental integral pipework, at the wellheads of the successful bores.

Collection pipework costs:

> The supply and installation of a suitable pipework system, with lagging, drainage and expansion facilities, for the purpose of collecting the geothermal fluids from the wellheads and delivering them to the point of need.

It may be noted that some of these activities will overlap in time. Exploration costs may continue more or less indefinitely, though at a reducing tempo, even after a field has been initially put to use. Drilling costs may also proceed for a long time, and wellheads and pipework may be added as development grows.

2.2 RECURRING COSTS

Interest on capital expenditure.
Depreciation of capital assets or amortisation of loans.

Maintenance and repairs of pipework, valves and wellhead equipment, and possibly the maintenance of bores by descaling or the rectification of fractures, etc.

Bore replacements, which may be necessary from time to time to make good the loss of fluid yields from ageing bores.

Salaries and wages of the operating, inspecting and supervisory staff.

All these recurring costs are fixed, and totally independent of the quantity of heat yielded by the field and delivered to the point of use.

3. Basic assumptions

Of equal importance to the values assigned to the various cost components (which together will determine the numerator of the heat costs) is the making of realistic assumptions as to the quantity of heat produced by a field (which determines the denominator of the heat costs). Fields of course differ greatly one from another, but if any idea is to be obtained of the order of magnitude of geothermal heat costs (in U.S. cents per million BTU) it is necessary to assign realistic values to the number of productive bores in the field and the heat output per bore. This can only be done in the light of general experience to date, and the following assumptions are suggested:

3.1 NUMBER OF PRODUCTION BORES IN A FIELD

The cost of heat will clearly depend upon the number of successful production bores over which the exploration costs may be spread. At Wairakei in New Zealand there are more than sixty production bores; at the Geysers in California there are about the same number; at Larderello in Tuscany there are more than two hundred. It would be a cautious assumption, implying only a moderate scale of development, to assume a total of fifty bores (successfully productive) in the hypothetical field under consideration.

3.2 HEAT YIELD PER BORE AND OPERATING PRESSURE

These two parameters should be considered together because they are interdependent. Figure 1 shows the average characteristics of various numbers of bores at each of five wet fields—Wairakei and Broadlands in New Zealand, Hveragerdi in Iceland, Otake in Japan and Mexicali in Mexico. The number of bores taken at each field is determined by the availability of data, and the figures represent measurements taken fairly early in the historical developments of the fields. Allowing for the tendency of steam yields to decline with time, and to become rather drier, and taking into consideration the modern trend to adopt rather moderate wellhead pressures so as to prolong the life of fields, a reasonable yield for a 'typical' bore might be taken at 100,000 lb/h of steam and 250,000 lb/h of hot water at a wellhead pressure of 75 psig. Single point yields typical of the dry field at Larderello in Italy is also shown

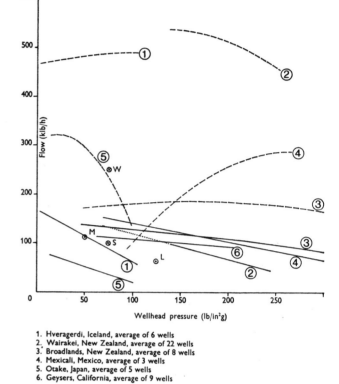

1. Hveragerdi, Iceland, average of 6 wells
2. Wairakei, New Zealand, average of 22 wells
3. Broadlands, New Zealand, average of 8 wells
4. Mexicali, Mexico, average of 3 wells
5. Otake, Japan, average of 5 wells
6. Geysers, California, average of 9 wells

L. Larderello	Steam	Assumed steam and water ⊗ S	—— Steam
M. Matsukawa		Outputs of 'typical' well ⦿ W	----- Hot water

FIG. 1. Collected well characteristics.

in Figure 1, but as wet fields are more common than dry the former have been given predominant consideration.

As a first approximation, the heat yield from a typical bore may therefore be taken as follows:

	Flow lb/h		Enthalpy BTU/lb		Heat Flow BTU/h
Steam	100,000	×	1,186.1	=	118,610,000
Water	250,000	×	290.4	=	72,600,000
Total	350,000	×	546.3 (mean)	=	191,210,000

These figures of enthalpy, being taken from the steam tables, are reckoned above 32 °F. It would be more reasonable to assess the heat of the bores as above that of water at ambient temperature, say 67 °F. It is thus necessary to deduct a figure of 35 (i.e. 67-32) from the above enthalpies in order to arrive at a more realistic estimate of the net heat content of the bore fluid, so:

	Flow lb/h		Enthalpy BTU/lb		Heat Flow BTU/h
Steam	100,000	×	1,151.1	=	115,100,000
Water	250,000	×	255.4	=	63,850,000
Total	350,000	×	511.3 (mean)	=	178,950,000

4. Estimated theoretical capital costs

It is now possible to estimate the theoretical capital expenditure in a 'typical' developed field, so:

4.1 EXPLORATION COSTS

It is of course possible to spend almost anything on geothermal exploration, but it is worth noting that three geothermal exploration projects in the Middle East and in Latin America have been financed in recent years by the United Nations Development Programme Special Fund, and that the costs of these have averaged rather less than $3 millions each and have covered a wide range of investigations and a moderate amount of exploratory drilling. A further exploration planned, but not yet executed, in the Far East has been estimated at $2½ millions. In a study performed in September, 1964, G. Facca and A. Ten Dam estimated typical exploration costs at figures ranging from $872,000 to $3,153,000 according to the scale of operations (Facca and Ten Dam, 1964). A figure of *$3 millions* may therefore be regarded as a conservative estimate for geothermal exploration. It will further be assumed that the whole of this sum is borne by a single developed zone. Spreading this cost over the assumed number (50) of bores, the exploration costs may thus be expressed at *$60,000 per bore.*

4.2 DRILLING COSTS

These must of course depend upon the nature of the ground drilled, on its location, on the depth of the bores and on the scale of operations. Bearing in mind that the average cost of oil wells in recent years, of average depth of 3,854 ft, has been just over $50,000 per bore, that the depth of geothermal bores is usually about 2,500 ft and that geothermal drilling is usually rather costlier than oil drilling, a cost per well of $60,000 would be a cautious figure. Assuming further that of every three bores sunk only two are successful and the third is a failure (in California and New Zealand a much higher success ratio has been achieved), the cost *per successful bore* may reasonably be taken at 3/2 × 60,000 or *$90,000.*

4.3 WELLHEAD EQUIPMENT COSTS

To cover the cost of a separator, silencer, valving, integral pipework and a reasonable degree of instrumentation (pressure gauges, thermometers and flowmeters) an average figure of, say *$35,000* per bore would be reasonable.

4.4 COLLECTION PIPEWORK COSTS

This is a very difficult figure to estimate rationally, as it can vary so enormously from field to field. But first let it be assumed that the mean distance from the bores to the point of delivery is *one mile* (i.e. further for the remote bores and less for the nearer ones). Next, combining the assumed well characteristics with the assumed number of bores, the total quantities of fluid handled will be:

Steam	50 × 100,000 = 5,000,000 lb/h
Water	50 × 250,000 = 12,500,000 lb/h

Assuming a maximum steam velocity of 150 ft/sec and a delivered steam pressure of 60 psig, the 5,000,000 lb/h of steam could be carried by eleven pipes of 30-inch internal diameter each. Pipework of this size, with lagging, expansion facilities, traps and suitable valving can be supplied and erected at a cost of about $90 per foot. Hence the cost of the steam mains may be estimated at:

11 pipes × 5,280 ft × $90 =	$5,227,200
Allow 40% for branch lines	$2,090,880
Total	$7,318,080
say	$7,500,000

If the hot water is also transmitted to the same point of delivery it could be accommodated in pipes at a speed of about 10 ft/sec if it is pressurised and/or attempered. The 12,500,000 lb/h of hot water would require three pipes of 20-inch internal diameter each. Pipework of this size can be supplied and erected at about $120 per foot. Hence the cost of the hot water mains would be about:

3 pipes × 5,280 ft × $120 =	$1,900,800
Allow 40% for branch lines	$ 760,320
Total	$2,661,120
say	$2,700,000

To this must be added the costs of the terminal equipment such as pumps, head tank, flashing equipment and control gear which, based on the experience of the Wairakei experimental hot water transmission, may be valued at about $3 millions. Hence the total hot water transmission costs may be estimated at:

Pipelines and branches	$2,700,000
Terminal equipment	$3,000,000
Total	$5,700,000

Adding to this the cost of the steam transmission a total heat transmission cost is arrived at of *$13,200,000,* or *$264,000 per bore.*

5. Theoretical estimated recurring costs

These too may be estimated as follows:

5.1 CAPITAL CHARGES

Interest charges will be taken at various rates, as the heat costs are sensitive to the rate chosen. Depreciation or amortisation costs may then be estimated on a sinking fund basis for each rate of interest according to the following assumed lives of the assets:

Bores	10 years
Wellhead gear and collection pipework	25 years

A word of justification of these assumed lives would not be out of place. There is usually a tendency for a bore's output of heat to decline in the course of time at rates which vary from field to field but which may be of the order of 5% to 10% per annum. The assumption in para. 3.2 of typical fluid yields is intended to be applied to average figures during the life of each bore: initially the yields could be expected to be higher, and ultimately they would be lower. Occasionally a bore may fall out of action altogether as a result of changes in the aquifer, persistent blockage or other cause. Although bores should generally remain productive for much longer than ten years it would be a cautious assumption to assign this life to them. Wellhead equipment and collection pipework on the other hand should be capable of extremely long lives, but a conservative figure of 25 years is proposed simply because the life of the field itself is to some extent uncertain, and it would be unwise to assume a longer life to these hardware assets than to the field which they serve, or that the field has a longer life than the utilisation plant which it feeds. Based upon experience in Italy, New Zealand and elsewhere 25 years would seem to be a very conservative figure. For the same reason it would be wise to assume that the costs of exploration are amortised over twenty-five years.

5.2 BORE REPLACEMENT COSTS

In assigning a life of only ten years to the bores the actual cost of the bore replacements as such would be covered by the sinking fund provisions. But whenever a bore falls out of use and has to be replaced by another in order to maintain the required total heat yield, certain incidental expenses are incurred. Although the same separator, silencer (pump), and instrumentation may be transferred to the site of the new replacement bore, costs will be incurred in moving these pieces of equipment and in modifying the shape of the pipework to conform with the configuration of the new bore in relation to the steam and water mains. It would be advisable to allow a figure of, say *$15,000* for this necessary adaptation per new bore.

5.3 OPERATION, REPAIRS AND MAINTENANCE

An allowance of *2% per annum* would be a reasonable figure for covering attendance and the repairs and maintenance of wellhead equipment, pipework, pumping and control equipment and instrumentation, etc. The actual bores themselves should not theoretically require any repairs or maintenance; but nevertheless blockages may sometimes occur which must be cleared, and casings can sometimes become damaged and must be repaired. A similar allowance of *2% per annum* would be reasonable for repairing such mishaps and for routine inspection of bores.

6. Cost of heat at the bores

On all these assumptions it is now possible to assign an approximate value to heat as it issues out of bores in the field. The supporting calculations are detailed in Table 1 for various interest rates, and the results are plotted graphically in Figure 2. Two sets of figures have been deduced, one for 100% and one for 90% load factor. The former is based on the assumption that the heat flows continuously

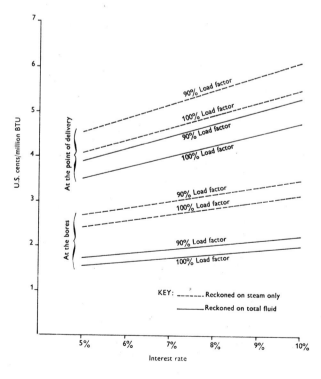

FIG. 2. Theoretical heat costs.

TABLE 1. Theoretical cost of heat at the bores

Interest rate			5%	6%	7%	8%	10%
			$p.a.	$p.a.	$p.a.	$p.a.	$p.a.
1. *Interest on:*							
	Exploration	$ 60,000/bore					
	Drilling	$ 90,000/bore					
	Wellhead equipment	$ 35,000/bore					
	Total	$185,000/bore	9,250	11,100	12,950	14,800	18,500
2. *Sinking Fund:*							
on 25-year basis on exploration and well- head equipment		$ 95,000/bore	1,990	1,732	1,502	1,300	965
on 10-year basis {	on drilling costs	$ 90,000/bore					
	on bore replacement costs	$ 15,000/bore					
		$105,000/bore	8,350	7,960	7,595	7,255	6,590
3. *Operation, repairs and maintenance:* at 2% on wellhead equipment and drilling costs		$125,000/bore	2,500	2,500	2,500	2,500	2,500
4. Net total			22,090	23,292	24,547	25,855	28,555
5. Add 10% for *contingencies and management* costs			2,209	2,329	2,455	2,586	2,856
Grand total			24,299	25,621	27,002	28,441	31,411
			¢/10⁶ BTU	¢/10⁶ BTU	¢/10⁶ BTU	¢/10⁶ BTU	¢/10⁶ BTU
6. *Cost of heat,* based on average yields per bore of:							
6.1 at 100% load factor {	Steam	115,100,000 BTU/h	2.41	2.54	2.68	2.82	3.12
	Water	63,850,000 BTU/h	—	—	—	—	—
	Total	178,950,000 BTU/h	1.55	1.63	1.72	1.82	2.00
6.2 at 90% load factor {	Steam only		2.68	2.82	2.98	3.14	3.47
	Total fluid		1.72	1.81	1.91	2.02	2.22

TABLE 2. Theoretical cost of heat at point of delivery

Interest		5%	6%	7%	8%	10%
		$p.a.	$p.a.	$p.a.	$p.a.	$p.a.
1. *Cost at bores* (b/f from Table 1)		24,299	25,621	27,002	28,441	31,411
2. *Interest* on collection pipework and associated equipment	$13.2 millions, or $264,000/bore	13,200	15,840	18,480	21,120	26,400
3. *Sinking Fund* on 25-year basis on	$264,000/bore	5,530	4,810	4,180	3,610	2,685
4. *Operation, repairs and maintenance* at 2% p.a. on	$264,000/bore	5,280	5,280	5,280	5,280	5,280
5. Add 10% of items 2, 3 and 4 for *contingencies and management*		2,401	2,593	2,794	3,001	3,437
Total		50,710	54,144	57,736	61,452	69,213

Assume that $7\frac{1}{2}$% heat losses are incurred between the bores and the point of delivery.

		¢/10⁶ BTU	¢/10⁶ BTU	¢/10⁶ BTU	¢/10⁶ BTU	¢/10⁶ BTU
6. *Cost of heat:*						
6.1 at 100% load factor {	Steam only [1]	4.06	4.32	4.60	4.88	5.46
	Total fluid	3.50	3.73	3.98	4.24	4.77
6.2 at 90% load factor {	Steam only [1]	4.52	4.80	5.11	5.42	6.07
	Total fluid	3.89	4.15	4.43	4.72	5.30

1. Costs of water transmission excluded.

from a bore and can be used without interruption throughout the year. The latter assumes that for 10% of the time each bore is taken out of service, either through some defect of the bore itself or of the plant and equipment it serves. Experience shows that, for power generation at any rate, a bore load factor of 90% is quite realistic. The costs have also been deduced on the alternative assumptions that only the steam is used and that the heat of the total fluid is used.

7. Cost of heat at collective point of delivery

By adding to the heat costs at the bores the capital charges and other costs arising from the fluid collection and transmission system, as deduced in Table 2, it is possible to estimate the cost of heat at the point of delivery of the transmission mains—both on the basis of steam only and of total fluid, and also at 100% and 90% load factor, as before. In this reckoning it is assumed that $7\frac{1}{2}$% of the heat is lost in transmission; largely apparent as steam condensation. This is a conservative allowance. The results are also plotted in Figure 2.

8. Commentary on theoretical heat costs

It is suggested that only the figures for 90% load factor be taken into consideration, the 100% figure being too idealistic. It will be observed from Figure 2 that the costs are not quite so sensitive to interest rate as might have been expected, though an influence is of course clearly marked. The figures show that if the heat of all the bore fluid is used it should be possible to obtain geothermal heat at a cost of the order of U.S. cents 2 per million BTU at the wellheads and at about U.S. cents 4 or 5 per million BTU if the heat is collected and piped to a single point about a mile away. If only the steam is used, these costs would be increased by about 50% at the bores and about 15% at the point of delivery.

These theoretical costs are based upon certain assumptions which, on the whole, have been cautiously chosen Although in some fields the heat costs would be higher· it is possible that in others they could be lower.

9. Actual heat costs

It is now of interest to examine such heat costs as actually occur in existing exploited geothermal fields. Very little information relating directly to heat costs has been published, but in some cases these costs can be indirectly deduced.

9.1 LARDERELLO

From indirect evidence published by Facca and Ten Dam (1964) it can be deduced that the cost of steam heat at the bores is probably about U.S. cents 2.9 per million BTU reckoned at an interest rate of 6%. If adjusted for interest at 8%, this cost would rise to about U.S. cents 3.2 per million BTU. These figures conform very closely with Figure 2.

9.2 ROTORUA

Heat costs at the wellhead at Rotorua in New Zealand have been quoted at U.S. cents 3.36 per million BTU in a paper describing the use of geothermal heat for air conditioning (Reynolds, 1970), but the assumed rate of interest, bore life and load factor are not specified. The scale of development is very small. By comparison, a figure of U.S. cents 67 per million BTU is given for oil fuel heat at the same place, which shows how very cheap geothermal heat can relatively be.

9.3 ICELAND

Heat costs for space heating purposes have been quoted at figures ranging from U.S. cents 7.5 to U.S. cents 12.1 per million BTU in the thermal area (Bodvarsson and Zoëga, 1964), but the basic parameters have not been given. Although these figures are considerably higher than the theoretical costs deduced earlier they are for a smaller scale of development. These costs rise to about U.S. cents 13 or 14 per million BTU if reckoned at the point of bulk delivery in Reykjavik City, and to U.S. cents 102 per million BTU if reckoned at consumers' premises after bearing the costs of the heat reticulation system. This last figure is nevertheless only 58% of the cost of oil fuel heat at consumers' premises. Elsewhere in Iceland, in the Námafjall area where there is a small scale industrial and heating installation, costs of U.S. cents 9 to 13.6 per million BTU have been quoted (Ragnars et al., 1970).

9.4 WAIRAKEI

On a basis of 8% interest, 90% load factor, a ten-year bore life and a 25-year life for other assets it can be shown that the cost of heat in the total fluid at the bores is about U.S. cents 2.65 per million BTU, by comparison with U.S. cents 2 as revealed by Figure 2. The difference can be accounted for partly by the relatively high drilling and exploration costs in New Zealand, and partly by the high wetness, and therefore the low enthalpy of the bores. If the costs are reckoned in terms of the steam fraction only they work out at U.S. cents 6.9 per million BTU, as against the figure of U.S. cents 3.2 derived from Figure 2. This difference too is largely due to the wetness of the field. A similar discrepancy between theory and practice may be observed when considering the cost of steam heat delivered at the plant, which is estimated at about U.S. cents 13.3

per million BTU as against a theoretical figure of U.S. cents 5.5 per million BTU as derived from Figure 2.

9.5 THE GEYSERS, CALIFORNIA

At this plant it is only possible to estimate the *price* of heat, as distinct from the *cost*, because the steam is retailed in bulk by a commercial supplier to the users, The Pacific Gas and Power Company. From the price of $2\frac{1}{2}$ mils per kWh charged for steam it is a simple matter to deduce that the price of steam heat is about U.S. cents $12\frac{1}{2}$ per

million BTU delivered to the plant, which is rather less than the equivalent figure for Wairakei. This price, however, includes the trading profit of the suppliers.

Although most of these heat costs are substantially higher than the theoretical costs deduced earlier, it must be remembered that they are mostly for early developments, and in some cases on a small scale. With new large scale developments, there is no reason why costs should not be obtained fairly close to the theoretical.

The actual and theoretical costs are summarised in Table 3.

TABLE 3. Approximate costs of geothermal heat in practice and theory (U.S. cents per million BTU)

Basis		Larderello (power)	Rotorua (air conditioning)	Iceland (district heating)	Wairakei (power)	Geysers (power)	Theoretical[5]
Cost at wellheads	steam only	} 3.2[1]	} —	—	6.90	} N.A.	3.3
	total fluid		3.36	N.A.	2.65		2.0
Cost at single collection point in thermal field		—	—	7.5 to 12.1	—	—	—
Cost at single delivery point		N.A.	—	13 to 14[2] }	13.3[4]	12.5[6]	7.31[4]
Cost at point of use		N.A.	—	102[3] }			

1. With a dry field the cost of steam and of total fluid are of course the same.
2. Cost delivered in bulk in Reykjavik, excluding distribution to local consumers.
3. Average cost delivered to domestic consumers.
4. Steam heat only. Delivered at power station.

5. Wet field assumed.
6. Purchase price of steam.
N.A. Not available.
Interest is taken at 8% p.a. for Larderello, Wairakei and Theoretical. Elsewhere, the rate of interest has not been specified.

10. Fuel heat costs

For all these costs of geothermal heat to be meaningful, it is necessary to form some idea of what the costs of fuel-derived heat are likely to be.

Solid and liquid fuels are usually sold gravimetrically; gaseous fuels are sold volumetrically corrected to standard temperature and pressure. The following formulae may be used for converting the fuel price into the heat price:

For solid and liquid fuels:

$$C = 44,650 \frac{P}{H}$$

where P is the fuel price in U.S. dollars per long ton,
H is the calorific value of the fuel in BTU/lb,
C is the heat price in U.S. cents per million BTU.

For gaseous fuels:

$$C = 1,000 \frac{P}{H}$$

where P is the fuel price in U.S. cents per thousand cubic feet,
H is the calorific value of the fuel in BTU per cubic foot,
C is the heat price in U.S. cents per million BTU.

These two formulae are shown graphically in Figures 3 and 4, on which are marked various typical fuels as they occur in nature; that is to say, without being dried. (For example, dried peats have calorific values of 8,000 to 9,000 BTU/lb, but owing to their high moisture content undried peats have calorific values of 1,000 to 3,000 BTU/lb only.) It should be noted that in the formulae quoted above, and in Figures 3 and 4, it matters not whether the higher or lower calorific value be taken so long as the resulting price of heat is reckoned on the same basis.

The prices of heat in the form of fossil fuels can of course vary enormously from place to place, largely on account of differences of transportation costs and of taxation incidence. A range of U.S. cents 25 to 65, however, would be fairly representative of coals and heavy fuel oils in the more populous parts of the world.

For some purposes—e.g. space heating and certain industrial processes—heat in the form of hot water is quite acceptable, or even necessary. For other purposes it may be necessary to have heat in the form of steam. But in some cases—e.g. internal combustion engines, gas turbines, and certain industries requiring 'fire' heat—it is essential that the heat be supplied in the form of fuel. Without taking proper account of these differences of form, according to the use to which the heat is to be put, any direct comparison between the costs of alternative heat sources is apt to be misleading.

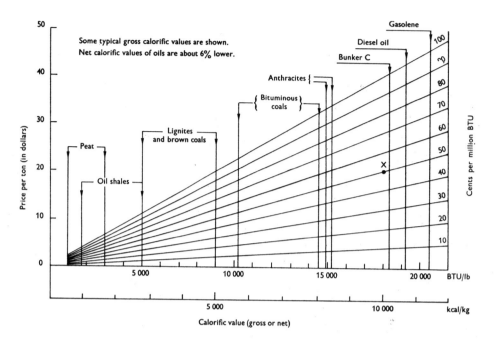

FIG. 3. Fuel and heat prices for solid and liquid fuels.

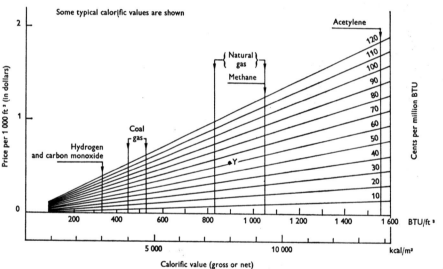

FIG. 4. Fuel and heat prices for gaseous fuels.

For instance, in the case of an industry requiring 'fire' heat (e.g. pottery) it is only the cost of fuel heat that is fully significant, since neither steam nor hot water could serve the required purpose. But where heat is needed in the form of steam or hot water, whether for a turbine or for industrial or domestic purposes, it is necessary to take into account the losses, the capital charges and the operating costs of producing the steam or hot water from the fuel. These costs will be considerably influenced by the scale of operations; but the approximate costs of heat in the form of steam, raised in boilers with fuel ranging in prices equivalent to U.S. cents 25 to 65 per million BTU, are shown in Table 4. The cost of heat in the form of hot water raised from fuel should be of the same order as for

steam, but clearly the scale would usually be nearer the 'small' end than the 'large', since hot water for any known industrial purpose at present is not likely to be required on a scale equivalent to a 500 MW power plant. (Even for the heating of the City of Reykjavik the total heat distributed is on a smaller scale than this.) Domestic heat on a very small scale, in the form of hot water produced from an individual fuel fired installation for a single household, may cost considerably more than even the highest figure in Table 4. For example, the average cost of such heat in Reykjavik has been quoted at U.S. cents 175 per million BTU (Bodvarsson and Zoëga, 1964).

There is sometimes a degree of flexibility of choice as to the form in which heat can be used. For instance, a

TABLE 4. Approximate costs of steam heat raised by fossil fuels (in U.S. cents/million BTU at point of use[1])

Scale of production (power plant assumed)	Cost of losses, of capital charges, and of operation of steam-raising plant, etc.	Assumed cost of heat in fuel	Total cost of heat in steam
Small (1 to 2 MW)	28-38	25-65	53-103
Large (500 MW)	13-18	25-65	38-83

1. e.g. turbine stop valve.

building can be heated equally well by steam as by hot water, and a similar freedom of choice would apply to many industries requiring process heat. Even in the case of power generation there can be some such flexibility, for it is possible to generate power from hot water as well as from steam and from the internal combustion of fuel. But in the case of power generation it is necessary to take proper account also of the grade of heat when comparing the costs of alternative heat sources.

The grade of heat largely determines the efficiency at which it can be used for power generation, and it is for this reason that alternative heat costs must be weighed with discretion. For example, it will be seen from Table 3 that heat is delivered in bulk to the City of Reykjavik at a cost only slightly exceeding the price of steam delivered to the power plant at the Geysers, California; but in the former case the heat is of a lower grade, suitable for its intended purpose of district heating but less suited for economic power generation than the steam heat delivered to the Geysers plant. (This does not imply that the heat in the thermal areas of Iceland could not have been used economically for power generation rather than for district heating.) Also from Table 3 it would appear that the cost of steam heat at the wellheads is more than twice as much at Wairakei as at Larderello, but the true economic difference in value is less marked because the higher pressure and temperature of the Wairakei steam should enable rather cheaper and more efficient plant to be used. Likewise, and for the same reasons, the differences in steam heat costs for small and large plants, as shown in Table 4, are less than the differences in the worth.

11. Optimum use of geothermal bores

Where geothermal steam is to be used for power generation there will be an optimum working pressure at which the power potential of a bore will have a maximum value. This is because of the inverse relationship between steam yield and pressure. The economic optimum pressure will differ somewhat from this 'maximum power optimum' owing to various considerations. On the one hand the adoption of a higher pressure (and therefore a higher steam density)

results in economies in the costs of pressure vessels, pipework and turbines. On the other hand the adoption of a lower pressure will sometimes enable certain wells of secondary quality to be used whereas they would have to be abandoned if a higher pressure were adopted owing to inability to make the grade. Generally the 'economic optimum' pressure will be rather higher than the 'maximum power' optimum pressure.

Where geothermal fluid is required as heat only, these considerations must be modified, for the lower the working pressure the greater will be the total heat yield. So long as sufficient temperature is maintained for the process required it will generally pay to work at as low a pressure as possible; though considerations of steam density will also exert some economic influence, particularly where long transmission distances are involved.

The actual heat costs shown in Table 3 are based, in three cases, upon the use of the heat for power generation. If these fields were used for low grade process heat the wells could be used at much lower pressures and the heat costs would be considerably reduced owing to the increased yields that would result.

12. The costs of geothermal power: general

For the generation of power from geothermal energy it is necessary to adopt pressures and temperatures well below those that would be used for conventional steam plants. Accordingly it might be expected that geothermal power would be expensive, since low pressures and temperatures are not conducive to high efficiencies. However, the extreme cheapness of geothermal heat by comparison with fuel-raised steam heat, as revealed in Sections 9 and 10 above, can be such as to offset completely the technical disadvantages of the low steam conditions and to ensure that very cheap power costs can be obtained with geothermal heat.

As in the case of heat, it is possible to estimate a theoretical 'standard' for the cost of geothermal power production, which must start with an estimation of the capital costs, and to see how this standard fits into the perspective of power production costs from other sources.

13. Theoretical capital costs of geothermal power

13.1 EXPLORATION COSTS

As before (para. 4.1) it will be assumed that the exploration costs amount to U.S. $3 millions.

13.2 DRILLING COSTS

As before (para. 4.2) it will be assumed that the cost of drilling each successful bore will be U.S. $90,000 after allowing for bore failures.

13.3 WELLHEAD GEAR AND FIELD PIPEWORK

As before (para. 4.3) it will be assumed that the cost of the wellhead equipment is U.S. $35,000 per bore. It will also be assumed as before (para. 4.4) that collection pipework costs U.S. $13,200,000 for a 50-bore installation; i.e. U.S. $264,000 per bore. Adding this to the costs of the wellhead equipment, the two together amount to U.S. $299,000 per bore.

13.4 POWER HOUSE PLANT, INCLUDING BUILDINGS AND COOLING WATER FACILITIES

Based on existing geothermal installations, the following costs may be taken as reasonable. It will be noted that they are less sensitive to scale than conventional thermal plants, owing to the need to use units of fairly moderate size even for large total installed capacities:

For 20 MW installed	U.S. $160/kW
For 50 MW installed	U.S. $140/kW
For 100 MW installed	U.S. $125/kW
For 200 MW installed	U.S. $110/kW

13.5 WELL CHARACTERISTICS

As before (para. 3.2) it will be assumed that a typical bore yields 100 klb/h of steam and 250 klb/h of hot water at a wellhead pressure of 75 psig. Assuming that the hot water is also used for power production (and to be consistent with para. 13.3 above which estimates the pipework costs including that required for hot water transmission), and taking an entry pressure to the turbine of 60 psig, the power potential of a typical bore would be approximately as follows:

From the steam	100 klb/h × 60 kWh/klb =	6,000 kW
From the hot water	250 klb/h × 4.2 kWh/klb =	1,050 kW
Total		7,050 kW

These figures assume a back-pressure of $2\frac{1}{2}$ inHg and the adoption of two-stage hot water flashing.

13.6 NUMBER AND COST OF PRODUCTION WELLS

It would be advisable to have a reserve of about 20% of the useful bores in order to ensure continuity of output, thus allowing for repairs to wellheads, blockages and other unforeseen happenings. The number of useful production bores required may therefore be assessed in the manner shown in Table 5.

13.7 TOTAL COSTS PER KILOWATT

In Table 5 are deduced the total capital costs per kilowatt for different total power installations on the basis of the various assumptions made above.

TABLE 5. Component and total kilowatt costs for geothermal power

Output	20 MW	50 MW	100 MW	200 MW
1. Theoretical number of wells $= \dfrac{\text{kW}}{7,050}$	2.84	7.10	14.20	28.40
2. 20% spare wells	0.57	1.42	2.84	5.68
3. Theoretical number of wells	3.41	8.52	17.04	34.08
4. Total practical number of wells	4	9	18	35
5. Drilling costs at $90,000/well	$360,000	$810,000	$1,620,000	$3,150,000
6. Drilling costs per kW	$18	$16.2	$16.2	$15.7
7. Exploration costs	$3,000,000	$3,000,000	$3,000,000	$3,000,000
8. Exploration costs per kW	$150	$60	$30	$15
9. Wellhead heat and collection pipework at $299,000/well	$1,196,000	$2,691,000	$5,382,000	$10,465,000
10. Wellhead gear and collection pipework costs per kW	$59.8	$53.8	$53.8	$52.4

	$/kW	$/kW	$/kW	$/kW
Cost summary:				
Exploration	150	60	30	15
Drilling	18	16.2	16.2	15.7
Wellhead gear and collection pipework	59.8	53.8	53.8	52.4
Plant, etc.	160	140	125	110
	387.8	270	225	193.1
Add 20% for interest during construction and contingencies	77.6	54	45	39.5
Total	465.4	324	270	232.6

14. Theoretical production costs for geothermal power

Based on the theoretical capital costs as assessed above, it is possible to deduce the theoretical production costs as follows:

14.1 ASSUMPTIONS

The following conservative assumptions will be made:
(a) That the plant will operate at an annual plant factor of 85%;
(b) That the useful life of the power plant be taken at 25 years, but that the life of the bores be taken at ten years only. This allows for replacing bores whose yields may have declined considerably or which may have become blocked or otherwise unusable;
(c) That every time a replacement bore is brought into service an expenditure of U.S. $15,000 is incurred in alterations to pipework and wellhead gear;
(d) That the rate of interest be taken at 8% p.a.

14.2 OPERATION, REPAIRS AND MAINTENANCE

Based on experience, the following figures may be taken as reasonable:

Installed capacity	20 MW	50 MW	100 MW	200 MW
Operations, repairs and maintenance in $p.a./kW	3.7	3.0	2.6	2.25

14.3 ALTERATIONS TO ACCOMMODATE REPLACEMENT BORES

Taking into consideration the number of bores required (see Table 5) and the assumed cost of alterations at U.S. $15,000 per bore, the annual costs of such alterations would be as follows:

Installed capacity	20 MW	50 MW	100 MW	200 MW
No. of bores in use	4	9	18	35
No. of replacement bores per annum assuming a 10-year bore life	0.4	0.9	1.8	3.5
Cost of alterations at $15,000 per bore	$6,000	$13,500	$27,000	$52,500
Equivalent in $ per kW per annum	0.3	0.27	0.27	0.2625

14.4 TOTAL PRODUCTION COSTS

The total production costs will therefore be as set out in Table 6. These costs are expressed both in dollars per annum per kilowatt and in mils per kilowatt-hour assuming an 85% load factor. The latter are shown plotted graphically in Figure 5. It will be seen that these costs compare very favourably with conventionally generated steam power.

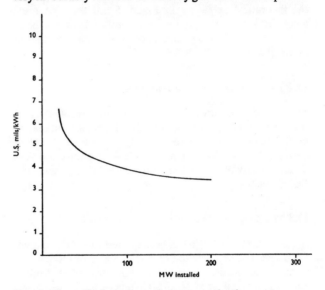

FIG. 5. Theoretical geothermal power production costs.

TABLE 6. Theoretical geothermal production costs

Installed capacity	20 MW	50 MW	100 MW	200 MW
	$p.a./kW	$p.a./kW	$p.a./kW	$p.a./kW
Interest at 8%[1]	37.23	25.92	21.6	18.61
Amortisation on sinking fund basis:				
Bores (10 years)[1]	1.49	1.34	1.34	1.30
Other assets (25 years)[1]	6.07	4.17	3.43	2.92
Operation, repairs and maintenance	3.70	3.00	2.60	2.25
Alterations to wellheads, etc.	0.30	0.27	0.27	0.26
Total	48.79	34.70	29.24	25.34
Equivalent cost/kWh in U.S. mils at 85% plant factor	6.55	4.66	3.93	3.41

1. Including 20% for interest during construction and contingencies.

15. Actual costs of geothermal power

Not very much data are available of geothermal power costs achieved in practice, and where costs have been quoted there has often been no mention of the assumed rate of interest or of the assumed asset lives. However, the following figures may be of some interest.

15.1 WAIRAKEI

Electrical energy is being generated at Wairakei (192 MW installed) at a cost, adjusted for 8% interest, and assuming a bore life of 10 years and a life of 25 years for other assets, of *5.14 mils/kWh*. This is nearly 50% more than the theoretical as shown in Figure 5 but the pipe route length is very long and the early evolution of the plant was complicated by a planned chemical plant which never materialised.

15.2 LARDERELLO (390 MW INSTALLED)

The cost of electricity is believed to be considerably cheaper than in New Zealand. Facca and Ten Dam have estimated, on an assumed interest rate of 6%, a generating cost of 2.6 to 2.74 mils/kWh for non-condensing plants and 2.38 to 2.96 mils/kWh for condensing plants (Facca and Ten Dam, 1964).

15.3 THE GEYSERS, CALIFORNIA (81 MW INSTALLED)

A cost of 4.81 mils/kWh has been quoted with fixed charges for return and depreciation at 8.57% p.a. Considering that in this case the steam has to be bought in bulk from a commercial supplier at $2\frac{1}{2}$ mils/kWh generated, this overall cost is astonishingly low.

15.4 NÁMAFJALL, ICELAND

For a small non-condensing set of only $2\frac{1}{2}$ MW production costs have been quoted at 'from $2\frac{1}{2}$ to $3\frac{1}{2}$ mils/kWh', but the underlying assumptions are not known (Ragnars *et al.*, 1970).

15.5 PAUZHETKA, KAMCHATKA, U.S.S.R.

An energy cost of only 7.2 mils/kWh has been quoted for this plant (Tikhonov and Dvorov, 1970), but without any underlying parameters being specified. Considering the small size of the plant (5 MW) and the very remote situation, this price can be regarded as low. It is claimed that the cost is 30% lower than could be achieved by an alternative means of power supply in the same area.

All these power costs are somewhat inconclusive, but they show that the theoretical costs derived earlier are not unrealistic.

16. Sundry economic aspects of geothermal power

It is axiomatic that geothermal power, the costs of which are fixed regardless of the quantity of energy generated, can attain minimum costs when used for base load purposes. Nevertheless, the author has elsewhere sought to show (Armstead, 1970) that it can be used economically for non-base load purposes. By designing a geothermal project to carry rather more than the system base load, the advantages of scale effect can sometimes offset the decline in load factor resulting from this. In fact for small systems it could sometimes be advantageous to carry the whole of the system load by means of a geothermal plant rather than the base load sector alone. This apparent paradox disappears when it is realised that after the system load has grown to such an extent that the base load overtakes the capacity of the geothermal plant, the geothermal production costs will fall to a minimum. Nevertheless, before this happens it can be advantageous to design a geothermal plant for initial working at somewhat reduced load factor. This can be a particularly attractive proposition if a geothermal plant is used in conjunction with special peaking plants of low fixed costs (e.g. gas turbines).

There can also be occasions where non-condensing geothermal plants can serve as peaking plants. This follows from their low capital cost per kilowatt—a necessary attribute for peaking plants. Such a practice, however, must be adopted with caution and confined to comparatively low outputs of energy, lest it should result in an intolerably rapid run-down of the thermal resources of the field.

Another aspect of geothermal power generation that may yet become important is in the use of secondary fluids (e.g. freons) as working media instead of water. The attractions of doing this are that a greater proportion of the available heat can be extracted from geothermal fluids—particularly if at relatively low temperatures—by means of heat exchangers, and that physically smaller turbines can be used by virtue of the relatively high vapour pressures of the secondary fluid. As against these advantages, however, costly heat exchangers and condensers have to be used, and special precautions have to be taken to avoid leakage of the secondary fluid to the atmosphere. Although some interesting experiments have been carried out in this direction, economic data are at present lacking (Pessina *et al.*, 1970; Moskvicheva, 1970).

17. Geothermal desalination

The use of geothermal heat for desalination purposes has not yet been put to the practical test, but the possibilities of doing this have been widely discussed. In a paper presented at the Dubrovnik International Symposium on Fresh Water from the Sea, September, 1970 (Armstead and Rhodes, 1970), the authors show that it should be possible to produce fresh water, under the right conditions, at the very low cost of about U.S. cents 29 per thousand U.S. gallons. Under very favourable conditions it should be possible to improve upon this figure.

18. Industrial and other applications of geothermal heat

Besides district heating, power generation and desalination, geothermal energy could be used for a great many heat consuming processes, of which the following is but an incomplete list:

Sugar processing in conjunction with paper manufacture from bagasse
Paper manufacturing from wood pulp
Total gasification of coal (Lurgi process)
Salt production
Borax production
Powdered coffee production
Dried milk production
Cattle meal from Bermuda grass
Rice parboiling
Textiles
Fruit or juice canning or bottling
Other food processing, canning or crop drying
Plastics
Timber seasoning
Fish drying and fish meal production
Recovery and processing of certain minerals (e.g. diatomite)
Refrigeration
Air conditioning
Horticulture and raising of vegetables under glass
Heavy water production
Recovery of valuable trace elements from geothermal waters.

Hitherto, not many of these applications have been put to practical use, but greenhouse heating is extensively practised in Iceland and the U.S.S.R., air conditioning is done in New Zealand, there is a very large paper manufactury in New Zealand, a diatomite recovery and processing plant has been established in Iceland, and plans have been drawn up for heavy water production in Iceland—all making use of geothermal energy.

The actual production costs of these industries are of little significance in the context of this article, but the point of importance is that these production costs are in every case lower than could be achieved by alternative means; and this arises from the fact that geothermal heat can be so very much cheaper than heat obtained from fuel. There is undoubtedly a wealth of potential for geothermal heat to be used for industrial purposes.

19. Multi-purpose plants

From what has been said in Sections 17 and 18 above it will be clear that the adoption of dual or multi-purpose geothermal plants could be economically advantageous. By combining power plants with other applications of geothermal energy it should be possible to share the costs of exploration, drilling and pipework between more than one end product. However, it should be remembered that the necessary restrictions of location of geothermal fields will often limit the choice of applications which can be combined at a single point. Clearly, for example, a desalination plant would be impractical at Wairakei where there is no shortage of water; nor would the raising of special crops in geothermally heated greenhouses be a suitable application in a tropical field such as might be encountered in the Philippines.

Certain heat consuming industries are also power-intensive, in which case it may be necessary to devote a large proportion of the power generated in a multi-purpose geothermal project for use within the industrial sector of the project. To quote an example, it is found that for geothermal desalination (Armstead and Rhodes, 1970) it should be possible to generate surplus power when desalting water so long as the performance ratio is relatively low, but if large water yields are required from a bore it may be necessary to import additional power. However, the potential of a good field may considerably exceed the steam requirements of a desalination plant, in which case there may be a surplus number of bores beyond those required for desalination, which can not only supply the power deficit for the desalting process but also provide a substantial margin of exportable power.

It is quite impossible to be specific in any general discussion of geothermal multi-purpose projects; each case must be examined on its own merits. But the point that requires emphasis is that whenever a geothermal field is being developed, serious consideration should always be given to applications other than power only. Ministries of Commerce and Industries should be asked to collaborate in seeking any heat consuming industry whatsoever that could profitably support the economy of the country in which the geothermal field occurs, so that a close study may be made of the possibilities of establishing geothermal multi-purpose projects.

20. General economic considerations

Apart from the fact that geothermal heat and geothermal power may be extremely cheap, there are other reasons why geothermal fields should be developed where possible. In the first place, even if its costs were not particularly favourable in themselves the use of a geothermal field enables an indigenous form of energy to be used, often with consequent savings of expenditure in foreign currencies on oil and other fuels. Although a geothermal power plant may be more capital-intensive than a conventional power plant, this factor may be of secondary importance by comparison with the enormous quantities of foreign currency that may be saved on the importation of fuels over the years.

Where industry is concerned, there are three location factors which govern the price of the end product—location of raw materials, location of the industrial process and location of the market for the end product. It may sometimes happen that it will pay to send raw materials to a geothermal field for processing, on account of the cheap heat, and to transport the manufactured products from that field to the market *even if the field is situated abroad*. There could be occasions where the logistics of doing this would involve less expenditure than the use of fuel heat close to the source of raw materials or to the markets. Thus the absence of a local geothermal field need not necessarily rule out the use of geothermal energy for heat-intensive industries.

21. Conclusion

Clearly geothermal energy cannot solve all energy problems. It is subject to limitations imposed by the accident of its location. There may also be occasions where the availability of a good geothermal field is nullified by formidable problems of the disposal of the waste fluids that cannot always be discharged into streams owing to a high content of such substances as boron, which can be highly poisonous to crops and to water supply systems. Nevertheless, geothermal heat can often be obtained without insoluble associated technical problems at costs that are far less than for any other form of heat, and of an acceptable grade. This fact can lead to cheap power, cheap industry and sometimes to large savings in foreign currencies.

Bibliography

ARMSTEAD, H. C. H. 1970. Geothermal power for non-base load purposes, United Nations Geothermal Symposium, Pisa, Italy. [1]

——; RHODES, C. 1970. Desalination by geothermal means. *International Symposium on Fresh Water from the Sea. Dubrovnik, Yugoslavia*, vol. 3, p. 451-459. Athens, European Federation of Chemical Engineering.

BODVARSSON, G.; ZOËGA, J. 1961. Heat production and distribution in Iceland. *Proc. U.N. Conf. on New Sources of Energy, Rome, Italy*, (paper G/37). New York, United Nations.

FACCA, G.; TEN DAM, A. 1964. *Geothermal power economics*, p. 8-10, 39. Los Angeles, Worldwide Geothermal Exploration Co.

MOSKVICHEVA, V. N. 1970. Geothermal power plant on the Paratunka River. United Nations Geothermal Symposium, Pisa, Italy.[1]

PESSINA, S. *et al.* 1970. Gravimetric loop for the generation of electric power from low temperature water. United Nations Geothermal Symposium, Pisa, Italy.[1]

RAGNARS, K. *et al.* 1970. Development of the Námafjall Area, Northern Iceland. United Nations Geothermal Symposium, Pisa, Italy.[1]

REYNOLDS, G. 1970. Cooling with geothermal heat. United Nations Geothermal Symposium, Pisa, Italy.[1]

TIKHONOV, A. N.; DVOROV, I. N. 1970. Development of research and utilisation of geothermal resources in the U.S.S.R. United Nations Geothermal Symposium, Pisa, Italy.[1]

1. Published by the Istituto Internazionale per le Ricerche Geotermiche, Lungarno Pacinotti 55, Pisa, Italy.

Management of a geothermal field

R. S. Bolton

Senior Design Engineer
Ministry of Works (New Zealand)

Estimation of the energy potential of a geothermal field

The development of a geothermal field is based on an estimate of its energy potential, together with estimates of the economics of development. Unfortunately, it is not yet possible to estimate the potential of a geothermal field with the same degree of precision as can be done for oil or gas fields. Essentially, the difference arises because in the case of hydrocarbons, the limits within which the energy is stored can be defined with reasonable accuracy using proven techniques. The nature of geothermal energy is such that it is extremely difficult to define these limits, the main problem being in determining depth. Nevertheless, estimates can be made which, by comparison with similar estimates for fields under exploitation, form a reasonable basis for evaluating the economics of development. This section discusses methods of estimating the potential, together with various related factors.

NATURAL HEAT FLOW

An assessment of the natural heat flow at the surface is usually one of the first steps undertaken in the investigations of a geothermal field. Basically, it represents a rate of withdrawal, and is therefore dependent primarily on the permeability of the system as a whole. It is useful in assessing the relative merits of different fields, (Bodvarsson, 1956; Grange, 1955). Also White (1965) suggests that it may be possible to produce a geothermal field at more than 5 times the rate of natural discharge, but points out that unless the deep inflow increases correspondingly, pressures and temperatures, and production rates will eventually decline. However, as an indication of the potential of a field, and indeed as a basis for comparison of different fields, the natural heat flow is not always reliable. In his comparison of the New Zealand fields, Grange shows the natural heat flow of the Broadlands field as about 2% of that of the Wairakei field. Subsequent investigations have shown the Broadlands field to have a potential comparable to that

of Wairakei. The discrepancy can be attributed partly to the difference in permeabilities of the two fields, but is also a reflection of the difficulty in obtaining a reliable assessment of the natural heat flow.

STORED HEAT

From a knowledge of the area of the field, of the distribution of temperature with depth and area, and of the formation characteristics, the heat stored in the formations between various levels can be estimated. By making assumptions leading to an estimate of the recoverable stored heat, the potential for power generation can be assessed. Estimates of this nature can be used as a basis for decision on further development, but must be regarded as qualitative rather than quantitative, and their limitations must be recognised.

The area of the field can be assessed on the basis of surface manifestations, such as surface heat flow, thermal alteration, geology, or by subsurface geophysical methods such as resistivity, gravity and magnetic anomaly surveys. Often, the resistivity surveys provide the most reliable indication of the area of the field. Subsurface temperatures of hot water systems may often be estimated from the chemistry of water discharged from hot springs (White, 1970), but the distribution of temperature both with area and depth can only be obtained by deep drilling. This also provides information on the formation porosity and permeability. Thus, the first limitation of this method of assessment is that it can only be made after a considerable amount of investigation work, including deep drilling, has been carried out.

A second, and more significant limitation is that, as already mentioned, it is very difficult, if not impossible to determine the effective depth of the system. In some fields, seismic surveys may permit an estimate to be made, as has been the case in Iceland. On the other hand, in New Zealand, seismic surveys give no indication of depth. Generally, in making estimates of stored heat, the depth taken is that proven by drilling, modified in some cases by a knowledge of the geology.

On the basis of chemical evidence, White (1970) states that at least 95% of the water recharging most geothermal systems is the local dominant meteoric water. Also, James (1970) concludes that hot water reservoirs can be regarded as 'once through' systems in which cold, meteoric water at the edge of the system passes deep down vertically, is heated, and rises to the surface in the hot area. It follows that a further limitation of the assessment of potential on the basis of heat stored in the upper levels is that it takes no account of the increase in flow through the system which will be induced by exploitation. This limitation is of course related to the depth limitation, and is also dependent on the permeability of the system as a whole which is the main factor determining the extent to which the flow through the system can be increased.

Thus, estimates of the potential based on stored heat should be regarded as minimum estimates of the actual potential. As such, they are useful in themselves when making decisions regarding development of the field, but their value in this direction is enhanced when used as a basis for comparison with similar estimates made for fields already under exploitation. Methods of assessing the potential on the basis of stored heat differ in details, and examples of two methods are given by Banwell and Bodvarsson (Banwell, 1964; Bodvarsson, 1970).

METHOD OF EXPLOITATION

It has been suggested (James, 1965*a* and *b*) that for a hot water reservoir, steam production from the upper levels of the production formation will result in the recovery of about twice the amount of energy as would water production from lower levels. This study presents another method of estimating the amount of energy which can be recovered from the stored heat. However, the main conclusion of the study as given above should not be accepted without a full understanding of the model used and of the assumptions on which it is based. There is some doubt as to the validity of one of the main assumptions, namely that the temperature/depth relationship in the water phase remains linear and constant throughout the exploitation period. If the alternative assumption is made that temperatures in the water phase follow a saturation temperature/pressure relationship with depth, the conclusion reached is that there is little difference in useful energy recovery between the two methods of exploitation. Also, deliberate attempts at production of steam from the upper levels of the main production aquifer at Wairakei were made in 1960 and again in 1963, but were unsuccessful. The two wells concerned were both wet, and had a small discharge. From a practical point of view therefore, it is unlikely that production of dry steam from the top of a production aquifer can be deliberately achieved, although due to local permeabilities and formation conditions isolated wells may so produce.

REINJECTION OF WASTE WATER

Reinjection of waste water from hot water reservoirs is not carried out in any fields at present being exploited. It is being seriously considered as a method of disposal of waste for the Ahuachapan field in El Salvador. There are several problems associated with reinjection, but from the point of view of recoverable energy, if it should prove to be practical, theoretical considerations show that it should also lead to a worthwhile improvement in the efficiency with which the total drawn off energy is used.

EFFECTS OF EXPLOITATION

In the case of oil and gas reservoirs, the principles of conservation of energy and mass have been developed into an extremely powerful tool in reservoir engineering, providing in many cases a reliable estimate of the energy initially in place. The application of this method is dependent on a period of exploitation which has had measurable effects on underground conditions in the field. Unfortunately, the development of these principles as they apply to geothermal reservoirs has not yet resulted in a similar method of analyses being available. One probable reason for this is that there are very few geothermal fields at present with any substantial production history.

The effects of exploitation are described later, but it is relevant to mention at this point that a simplified application of the mass balance principle to the Wairakei field indicated at a relatively early stage that inflow was having a significant effect. However, at present the most important application of a study of the effects of exploitation is in providing guidance as to the rate of development and to the maximum development the field will support. It follows that development should proceed in stages, the effects of each addition being studied before proceeding with the next.

Primary effects of exploitation

The number of geothermal fields under exploitation is increasing, but at present there are few with a production history indicating the effects of exploitation. The examples discussed below are the Wairakei field in New Zealand, the Larderello field in Italy and the Laugarnes field in Iceland. The first is basically a high temperature hot water system, the second a dry steam system and third a low temperature (less than 150 °C) hot water system. In this respect, they can be regarded as typical of the three main types of geothermal systems.

WAIRAKEI

The production history of this field extends over nearly 20 years, and in that time substantial changes have taken place in the underground conditions. Bolton (1970) gives a detailed description of the behaviour of the field under exploitation. The production aquifer is a breccia about 1,500 ft thick, overlain by a mudstone/siltstone formation which acts as a cap rock. Useful production is obtained from the breccia, but the best production is from the contact between the breccia and the underlying ignimbrite at a depth of about 2,000 ft. Initially, the aquifer was filled with water, temperature and pressure conditions following the boiling point for depth relationship. The maximum temperature measured in the field was 260 °C.

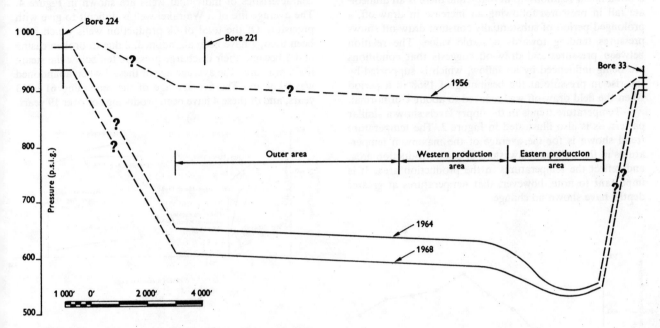

FIG. 1. Wairakei pressure profiles in 1956, 1964 and 1968 (Pressures at R.L.-900).

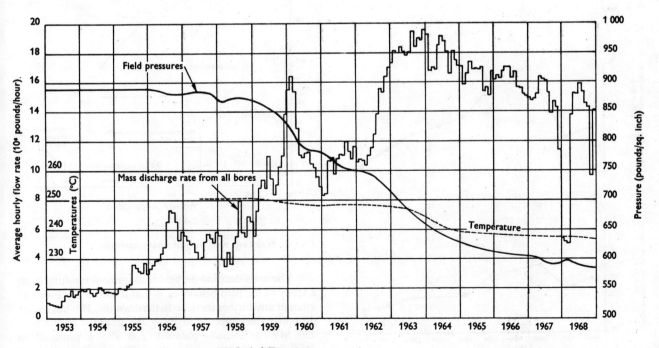

FIG. 2. Field discharge rate and pressures; Wairakei (Pressures shown are at R.L.-900).

Exploitation has resulted in an almost uniform pressure decline of over 300 psi and affecting an area considerably greater than the main production area (Fig. 1). This is indicative of the very high horizontal permeabilities in the field. Figure 2 shows the relationship between rate of pressure decline and rate of draw-off. As can be seen, this is not a linear relationship in that while there is an immediate fall in pressures following an increase in draw-off, a prolonged period of substantially constant draw-off shows pressures tending towards a stable value. The relation between pressures and draw-off suggests that conditions are being influenced by an inflow, which is supported by the rise in pressure at the beginning of 1968 in a period when the field draw-off was reduced to about $\frac{1}{3}$ of normal.

Temperature trends in the upper levels show a similar pattern as is also illustrated in Figure 2. The temperature trend shown is for the average of the maximum temperatures in the wells in the western or main production area, and reflect the temperatures in the production area. It is important to note, however, that temperatures at greater depths have shown no change.

Production trends are illustrated in Figures 2 and 3. Except for the first quarter of 1968, the number of wells on production has been substantially constant since 1963, but there has been a gradual decline in output. Figure 3 also shows the effect on total output, but expressed in terms of an average bore. Typical changes in the discharge characteristics of individual wells are shown in Figure 4. The average life of a Wairakei well is difficult to give with precision. Of the total of 68 production wells which have been used, 7 have been abandoned, 2 due to broken casing and 5 because their discharge pressure fell below the steam main pressure. The average age of these 7 when abandoned was $9\frac{1}{2}$ years. The average age of the remaining 61 is 13 years, and of these 4 have been producing for over 19 years.

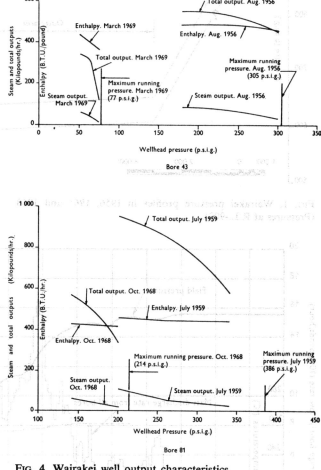

Fig. 4. Wairakei well output characteristics.

The main decision on field management resulting from a study of the field behaviour is that it would be unwise to attempt any further increase in the draw-off. This has meant dropping the proposal to bring the installed capacity to 250 MW as initially planned, and also a proposal to develop the outer area separately. The reason is that any substantial increase in draw-off will result in a further decline

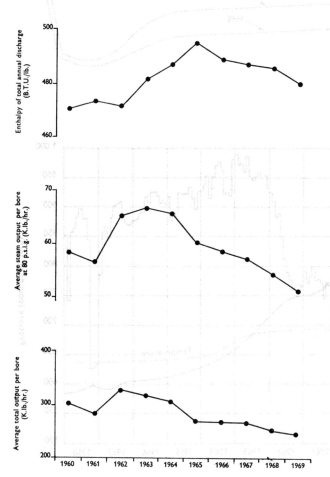

Fig. 3. Wairakei field discharge characteristics.

in pressures, temperatures and well output such that while there may be an immediate gain in total discharge, this would rapidly fall off, and there would be no long term benefit.

LARDERELLO

This field has been under exploitation for many years, initially for the recovery of chemicals, and since the mid 1920's, for the generation of electricity. As at Wairakei, substantial changes have taken place. It has not been the practice at Larderello to measure formation temperatures and pressures, but the nature of the field is such that changes in formation conditions can be inferred from wellhead measurements.

Figure 5 (Burgassi, 1964) shows the changes which occur in the output, and the wellhead pressures and temperatures in a typical well in the heavily exploited Larderello area. In lightly exploited areas, the output of a well will remain constant for ten years. The life of a steam well at Larderello is estimated to be 20 years (Chierici, 1964) but some are still discharging after 30 years, although at a greatly reduced output.

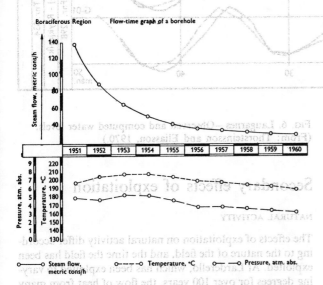

FIG. 5. Larderello well output characteristics. (From: Burgassi, 1964.)

The most significant change which has taken place in the Larderello field is that whereas the discharge from the initial deep bores was saturated steam, it is now dry with an appreciable amount of superheat. This is attributed by Chierici (1964), and Ferrara *et al.* (1970) and others to a gradually receding boiling water surface at depths in excess of 2 km, the steam from which is superheated by passing through layers of rock which, due to their poor conductivity, have retained heat.

A feature of the management of the Larderello field has been the need for a continued programme of drilling to maintain the total output, the newer wells being drilled relatively deeper. At present a number of production wells are operating close to their shut-in pressure, and exploration is continuing at still greater depths with the object of locating higher pressures.

LAUGARNES

This field has been under exploitation since 1928 for district heating in the city of Reykjavik. Initially, production was from shallow artesian wells, but since 1962 has been wholly by pumping from deep wells. The behaviour of this field, together with a comprehensive analysis covering the period 1957 to 1969 is described by Thorsteinsson and Eliasson (1970).

Production is from three aquifers. Temperatures in the upper aquifer, which is about 200 m below ground level are 110-120 °C. In the centre aquifer, the temperature is 135 °C and in the lower, about 2,200 m below ground level, 146 °C. From 1957 to 1962, withdrawal rates were relatively uniform as the flow was mainly artesian. Since 1962, following the introduction of deep well pumping, the flow has varied seasonally, the winter flow being about 3 times the summer flow. The winter peak draw-off is accompanied by a draw-down in aquifer pressures, but this draw-down is almost entirely recovered during the summer. From January 1957 to August 1969, the net decline in pressure was equivalent to 66.8 m waterhead. No effect on temperatures at depth has been reported.

Thorsteinsson and Eliasson have applied the principles of flow in porous media to the development of a linear relationship between draw-down and a time integral of pumping rate enabling them to match the response curve of the aquifer to a standard deviation of 2 m (Fig. 6). They have also found that the Theis non-equilibrium formula describes closely the initial response of the aquifer to sudden changes in pumping rate. As a result of their analysis, they conclude that providing there is no increase in pumping rates, there will be no significant increase in overall draw-down in the coming years.

SUMMARY OF THE PRIMARY EFFECTS OF EXPLOITATION

The substantial exploitation of any underground fluid will result in a decline in pressure of the system, and geothermal systems are no exception to this rule. In the case of fields discharging under thermo-artesian conditions, outputs also fall. The extent of the fall in output and pressures is governed by the rate of replacement of the discharged fluids, which in turn is a function of the permeability of the system as a whole. At present, the magnitude of these effects cannot be predicted from early investigations, and can only be assessed after a period of exploitation. This is the fundamental reason why a geothermal field must be developed in stages.

In thermo-artesian systems, the discharge capacity of individual wells decreases with increasing exploitation. This

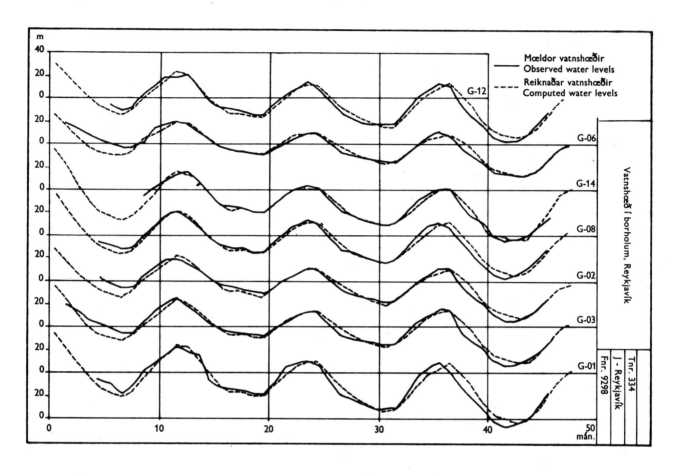

FIG. 6. Laugarnes—Observed and computed water levels.
(From: Thorsteinsson and Eliasson, 1970.)

indicates that when selecting an operating pressure, the lowest pressure consistent with the other factors influencing the choice should be adopted as this will give the longest operational life for the individual wells.

In a high temperature hot water system such as Wairakei, in which saturation conditions apply in the upper levels, the fall in pressure must inevitably be accompanied by a fall in temperatures at these levels. The heat stored in the formation rock will tend to sustain temperatures causing a lag between the temperatures and pressure declines. Similar effects are shown by a steam system, but in this case the effect on temperature will be less pronounced. There can, however, be a substantial increase in enthalpy.

With non-thermo artesian systems developed by deep well pumping, the effects of exploitation on output are obscured by the pumping, but there will nevertheless be a decline in pressures. In this type of system, where saturation pressures for the measured temperatures are considerably lower than the hydrostatic pressure, it is unlikely that steam will form in the formations. Consequently, the principles of flow in porous media can be successfully applied, as has been done for the Laugarnes field.

Secondary effects of exploitation

NATURAL ACTIVITY

The effects of exploitation on natural activity differ according to the nature of the field, and the time the field has been exploited. At Larderello, which has been exploited to varying degrees for over 100 years, the flow of heat from many of the areas of intense natural activity has almost disappeared (Elder, 1966). At Wairakei on the other hand, heat flow measurements in 1958 showed that the natural heat flow had not changed since 1952 although the field output was over $1\frac{1}{2}$ times the natural discharge (Ferrora et al., 1970). Subsequently, the field output has reached 6 times the natural discharge and, although no precise measurements are available, there is no noticeable decrease in the natural heat discharge. There has, however, been a noticeable increase in the enthalpy of the discharge. This is evident in the decline in or cessation of discharge from hot pools and springs, and an increase in area and intensity of heat escape from steaming ground, occasionally accompanied by small eruptions.

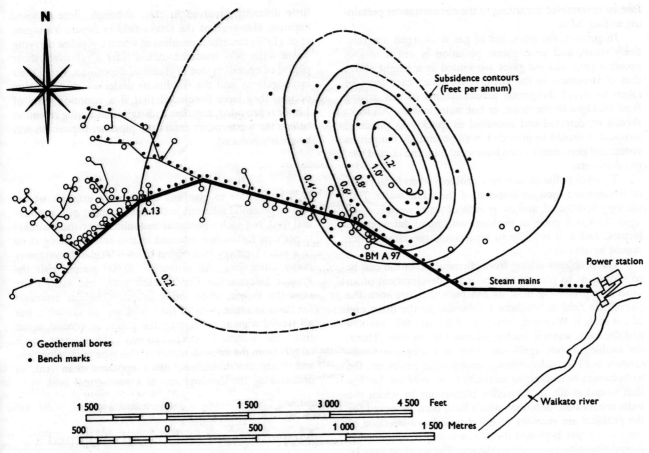

N

Subsidence contours
(Feet per annum)

1.2'
1.0'
0.8'
0.6'
0.40'
0.2'

A.13

● BM A 97

Power station

Steam mains

Waikato river

o Geothermal bores
● Bench marks

1 500   0   1 500   3 000   4 500   Feet

500   0   500   1 000   1 500   Metres

FIG. 7. Annual subsidence at Wairakei 1967.

GROUND MOVEMENT

Ground movement has occured after exploitation of many cold water aquifers and oil fields, the most well known examples of each probably being Mexico City, and Long Beach, California. The only similar effect so far reported for a geothermal field has been that at Wairakei (Hatton, 1970; Smith, 1967).

The movement at Wairakei has a strong vertical and horizontal component and has formed a roughly elliptical dish-shaped depression. The maximum rate of subsidence has been about 1.3 ft a year, with a total maximum subsidence estimated to be over 10 ft. The centre of the subsidence is about 1,500 ft from the nearest wells, and 6,000 ft from the region of greatest draw-off (Fig. 7). The movement is attributed to the bending of the mudstone cap rock resulting from the fall in pressure and the withdrawal of the large mass of fluid. The geology of the area of maximum subsidence is not known precisely, but there is a direct correlation between the subsidence and the thickness of the breccia underlying the mudstone in the region for which geological information is available. Precise measurements

taken by tiltmeter during the partial shut down in 1968 suggest that the subsidence may be reversible.

The ground movement has affected steam mains and drainage channels with some inconvenience, but without jeopardising the operation of the field. However, the effect of the movements measured at Wairakei on powerhouses or similar structures would be little short of disastrous. For this reason, the possibility of ground movement must always be considered, and a comprehensive system of bench marks installed in sufficient time to enable potential areas of subsidence to be located before the powerhouse site is committed.

POLLUTION

The main types of pollution likely to be encountered are atmospheric from the disposal of gas, thermal from the waste heat, and chemical from the dissolved salts in the waste water. One or more problems arising from the effects of pollution is likely to be encountered in any geothermal field. The extent and detailed nature of the problem will vary considerably for each field, and solutions must there-

181

fore be determined according to the circumstances pertaining to each field.

In general, the quantities of gas discharged are relatively small, and atmospheric pollution is not a serious problem provided the gases are vented at a height above that of structures in the vicinity. However, care must be taken to avoid dangerous accumulations of gas arising from leakages in the steam or hot water system. All leaks should be detected and remedied as quickly as possible; ventilation should be provided in all potentially dangerous areas; and personnel should have available, and should use, gas detectors.

Thermal pollution can be a problem when local rivers or streams are used for disposal of the waste water. Fish life may be affected, and the growth of water weed encouraged. This is a complex problem with many inter-related factors, and if it is likely to be serious, specialist advice should be obtained.

The problems arising from chemical pollution can be so serious as to prevent or hinder the development of an otherwise promising field, as has been the case with the Salton Sea field in Southern California. At the other end of the scale is Wairakei, where the dissolved salts are low and the waste water is discharged into a major river. There, the problem is not significant. Another example is Ahuachapan in El Salvador, where, among other problems, the wells cannot be discharged vertically because of the damage that would result to the coffee plantations in which the wells are located. Solutions which have been proposed for the problem are recovery of the chemicals, re-injection of the waste water both into the production field and outside it, and pumping the waste to the sea. The solution may be costly, but there may also be other economic or technical benefits which offset this cost to some degree.

Chemical pollution is normally a relatively minor problem in fields producing steam, but White (1970) mentions that pollution by boron and ammonia is now requiring preventative treatment at The Geysers, California. The solution adopted is the injection into the reservoir of the condensate containing the pollutants.

CHEMICAL DEPOSITION

Reduction of output due to chemical deposition in the wells has occurred in several fields, but removal of the deposits has been a straightforward process, and in most cases the output has been fully restored. One example is described by Cataldi et al. (1970). Deposition normally takes place at the level of boiling in the well which, as pressures decline due to exploitation, becomes progressively deeper. Ultimately, boiling and deposition will take place in the formations (Smith, 1967). At present, no evidence has been reported of chemical deposition in the formation having a detrimental effect: it can therefore be inferred that such deposition is a long term effect.

Chemical deposition can also be a problem in the disposal of waste water. At Wairakei, silica deposits in the waste water channel must be removed annually. There is little difficulty involved in this, although there is some expense. However, at the Otake field in Japan, Yanagase et al. (1970) describe a problem in which a pipeline carrying waste water was rendered useless after a relatively short period of operation due to chemical deposition. From their investigations into the conditions under which deposition occurs, they have concluded that if a retention time of 1 hour is provided, together with agitation during retention before the waste water enters the pipeline, deposition will be greatly reduced.

TOURISM

It is impossible to give any quantitative estimate of the tourist potential inherent in the development of a geothermal field, but such a potential undoubtedly exists. Wairakei is perhaps fortunately situated in this respect, being close to a main highway in a region of New Zealand containing many attractions. An estimated 30,000 people visit the Project Information Centre each year, and many more view the Project under other auspices. Similar estimates for fields in other parts of the world are not available, but all record a great interest by the public in general, apart from the technical interest shown. The interest stems largely from the unusual nature of this type of development and from the difficulties, more apparent than real, in undertaking the development of a geothermal field.

Measurements and records required

Except for low temperature geothermal fields, where the established theory of single phase flow in porous media can be applied, geothermal field management lacks the aid of the powerful mathematical tools available, for instance, in the management of oil and gas reservoirs. The lack of this ability is no barrier to the successful development of geothermal fields, but were it available, development could be controlled more rationally. Undoubtedly, as more geothermal fields are exploited, so will the theoretical and mathematical background expand. Notwithstanding this lack, the management of a geothermal field still requires the accumulation and assessment of data on which are based the decisions concerning the initial methods of exploitation, and the maintenance of or modifications to these methods as exploitation continues. The data are also essential in the development and testing of the relevant theory.

FIELD OUTPUT

The total mass and heat discharged from the field is fundamental both to field management and to development and testing of theory. For this purpose, a continuous record should be kept of the periods and rate of discharge for individual wells.

WELL OUTPUT CHARACTERISTICS

Periodic measurements of the mass and heat outputs over the range of possible wellhead pressures are necessary to check the variations which take place consequent on changes in field conditions.

WELLHEAD PRESSURE

This is of fundamental importance because it serves as a reference base when describing surface conditions, and in many cases when describing underground conditions also. It should be measured every time any measurement of any sort is made on a well.

WELLHEAD TEMPERATURE

In the case of high temperature hot water fields, this measurement is not so essential because, particularly when discharging, saturation conditions can reasonably be assumed and wellhead temperatures obtained from the measured pressure. For dry steam and low temperature hot water fields, this cannot be assumed, and in these cases, wellhead temperatures should be measured.

STATIC DOWNHOLE PRESSURES AND TEMPERATURES

These measurements, taken in a shut-in well in which conditions are relatively stable, are a direct indication of changes in underground conditions. As it is often impossible to ensure that truly stable conditions apply, considerable care is necessary when interpreting these measurements. To obtain a reliable indication of changes in field conditions, measurements should, in general, be taken at two to three monthly intervals in a number of wells, and in every well at least once a year. The frequency of measurement should also reflect the rate at which changes are occurring, and the degree of uniformity between measurements on different wells.

PRESSURES AND TEMPERATURES DURING DISCHARGE

These measurements, taken on individual wells, give information on feeding conditions to the well. For instance, measurements at Wairakei after some years of exploitation show that the temperature of the fluid feeding some of the wells differs from the formation temperature as measured in a static run. Also, pressure measurements during discharge may indicate whether permeability is changing. These measurements are not required regularly, but should be taken as the opportunity offers, particularly if changes of this nature are suspected.

GEOCHEMICAL MEASUREMENTS

Geochemistry is discussed elsewhere in this volume, but it should be mentioned here that the chemistry of the field discharge can provide much useful information on field behaviour, and regular monitoring is desirable.

GRAVITY MEASUREMENTS

Hunt (1970) describes a method not yet completely proven, but which looks very promising for assessing the net mass withdrawal from a field. This, by comparison with the total draw-off, gives an independent assessment of the extent to which inflow may be influencing conditions. The gravity measurements described for Wairakei suggest that inflow has a significant influence, providing confirmatory evidence for the conclusions drawn from the response of pressures to mass draw-off.

GROUND LEVEL MEASUREMENTS

These measurements are important early in the life of a field, preferably covering a period of substantial draw-off, and before sites are selected for the main structures. The frequency of subsequent measurements will depend on the extent and rate of movement, and the potential danger to steam mains or other works likely to be affected. Initially, measurements at about 2-yearly intervals should be sufficient.

ACCURACY AND INTERPRETATION OF MEASUREMENTS

Having undertaken a series of measurements, frequently with high quality instruments, there is always a strong tendency to accept without question the accuracy and reliability of the results. This tendency must always be guarded against, as there are many factors which can influence the accuracy. For this reason, every opportunity should be taken to make independent check measurements. For example, wellhead pressures can serve as a check on downhole pressure and temperature measurements, the former directly, the latter through the saturation pressure/temperature relationship if this applies. Again, the enthalpy from output tests can often, especially in the early days of exploitation, be checked against downhole temperatures. Any measurement of doubtful accuracy should be discarded, and a repeat measurement made.

Finally, when satisfied as to the accuracy of a particular measurement, or series of measurements, the results must be interpreted with a knowledge of all factors which could have influenced those measurements. This applies particularly when the results lead to unexpected conclusions, or conclusions at variance with past experience. Too often, neglect of this elementary principle can lead to erroneous conclusions.

Bibliography

Note. References to 'Rome' relate to the United Nations Conference on New Sources of Energy, 1961, the proceedings of which were published in 1964 by the United Nations, New York.

References to 'Pisa' relate to the United Nations Symposium on the Development and Utilisation of Geothermal Resources, 1970, the proceedings of which have been published by the Istituto Internazionale per le Ricerche Geotermiche, Lungarno Pacinotti 55, Pisa, Italy.

BANWELL, C. J. 1961. Geothermal Drill Holes—Physical Investigations. Rome, Paper G/53.

BODVARSSON, G. 1956. Natural Heat in Iceland. *Trans. 5th World Pwr Conf. Vienna.* (Paper 197K/8.)

——. 1970. An Estimate of the Natural Heat Resources in a Thermal Area in Iceland. Pisa.

BOLTON, R. S. 1970. The Behaviour of the Wairakei Geothermal Field During Exploitation. Pisa.

BURGASSI, R. 1961. Prospecting of Geothermal Fields and Exploration Necessary for their adequate Exploitation, performed in various regions of Italy. Rome, Paper G/65.

CATALDI, R.; ROSSI, A.; SQUARCI, P.; TAFFI, L.; STEFANI, G. 1970. Contribution to the knowledge of the Larderello Geothermal Region, Tuscany, Italy—Remarks on the Travale Field. Pisa.

CHIERICI, A. 1961. Planning of a Geothermal Power Plant: Technical and Economic Principles. Rome, Paper G/62.

ELDER, J. W. 1966. Heat and Mass Transfer in the Earth: Hydrothermal Systems. *Bull. N. Z. Dept. sci. industr. Res.,* 169.

FERRARA, G.; PANICHI, C.; STEFANI, G. 1970. Experimental Well in Larderello Steam Area—Remarks on the Geothermal Phenomenon in an Intensively Exploited Field. Pisa.

FISHER, R. G. 1964. Geothermal Heat Flow at Wairakei during 1958. *N.Z. J. Geol. Geophys.,* 7, 172-84.

GRANGE, L. I. 1955. Geothermal Steam for Power in New Zealand. *Bull. N.Z. sci. industr. Res.,* 117.

HATTON, J. 1970. Ground Subsidence of a Geothermal Field during Exploitation. Pisa.

HUNT, T. 1970. Net Mass Loss from the Wairakei Geothermal Field, New Zealand. Pisa.

JAMES, C. R. 1965a. Power Life of a Hydrothermal System. *Proc. 2nd Australasian Conf. on Hydraulics and Fluid Mechanics.* Auckland, New Zealand.

——. 1965b. Power life of a hydrothermal system. *International Symposium on Volcanology, New Zealand.* Rome, International association of volcanology and chemistry of the earth's interior.

——. 1970. Reservoir Physics and Production Management. Rapporteur's report, Section VII, Pisa.

SMITH, J. H. 1967. Exploitation of Geothermal Water. *Proc. Water for Peace Conf.,* Washington, vol. 3, p. 1. Superintendent of Documents, U.S. Government Printing Office.

THORSTEINSSON, T.; ELIASSON, J. 1970. Geohydrology of the Laugarnes Hydrothermal System in Reykjavik, Iceland. Pisa.

WHITE, D. E. 1965. Geothermal energy. *International Symposium on Volcanology, New Zealand.* International association of volcanology and chemistry of the earth's interior.

——. 1970. Geochemistry Applied to the Discovery, Evaluation and Exploitation of Geothermal Energy Resources. Rapporteur's report, Section V, Pisa.

YANAGASE, T.; SUGINOHARA, Y.; YANAGASE, K. 1970. The Properties of Scales and Methods to prevent them. Pisa.

List of authors' addresses

Mr. H. C. H. Armstead, Consulting Engineer,
Rock House,
Ridge Hill,
Dartmouth, South Devon
(United Kingdom)

Mr. John Banwell,
Consulting Geophysicist,
21 Putnam Street,
Wellington 5
(New Zealand)

Mr. R. S. Bolton,
Senior Design Engineer,
Ministry of Works,
P.O. Box 12-041,
Wellington
(New Zealand)

Mr. W. R. Braithwaite,
Dept. of Scientific and Industrial Research,
Chemistry Division,
Private Bag,
Petone
(New Zealand)

Sir Edward Bullard, F.R.S., Professor of Geophysics,
Department of Geodesy and Geophysics,
University of Cambridge,
Madingley Rise,
Madingley Road,
Cambridge. CB3 OEZ
(United Kingdom)

Mr. N. D. Dench,
Investigations Engineer,
Ministry of Works,
P.O. Box 12-041,
Wellington
(New Zealand)

Mr. Sveinn Einarsson,
Managing Director of 'Vermir' S.F.,
Research Engineers and Geophysicists,
Chairman of Directors of the Islandic
Institute of Industrial Research and Development,
Reykjavik
(Iceland)

Dr. Giancarlo Facca,
Consulting Geologist,
1023 Timothy Lane,
Lafayette,
California 94549
(U.S.A.)

Mr. Baldur Líndal,
Consulting Engineer,
Verkfraedistofa Baldurs Líndal,
Armúla 3,
Reykjavik
(Iceland)

Mr. T. Marshall,
Deputy Director,
Chemistry Division,
Department of Scientific
and Industrial Research,
Private Bag,
Petone
(New Zealand)

Mr. Keiji Matsuo,
President of the Teiseki Drilling Co.,
Japan Geothermal Energy Association,
Denki Kyokai Building 1-3,
Yuraku-Cho,
Chiyoda-ku,
Tokyo
(Japan)

Dr. James McNitt,
Project Manager,
Geothermal Resources Exploration Project,
P.O. Box 9368,
Nairobi
(Kenya)

Dr. Gudmundur E. Sigvaldason,
Science Institute,
University of Iceland,
3 Dunhaga,
Reykjavik
(Iceland)

Mr. J. H. Smith,
Chief Geothermal Engineer,
Ministry of Works,
P.O. Box 12-041,
Wellington
(New Zealand)

Mr. Basil Wood,
Associate of Merz and McLellan,
Consulting Engineers,
Milburn,
Esher, Surrey
(United Kingdom).